EVERY STEP IN
BEEKEEPING
BENJAMIN WALLACE DOUGLASS

EVERY STEP IN BEEKEEPING

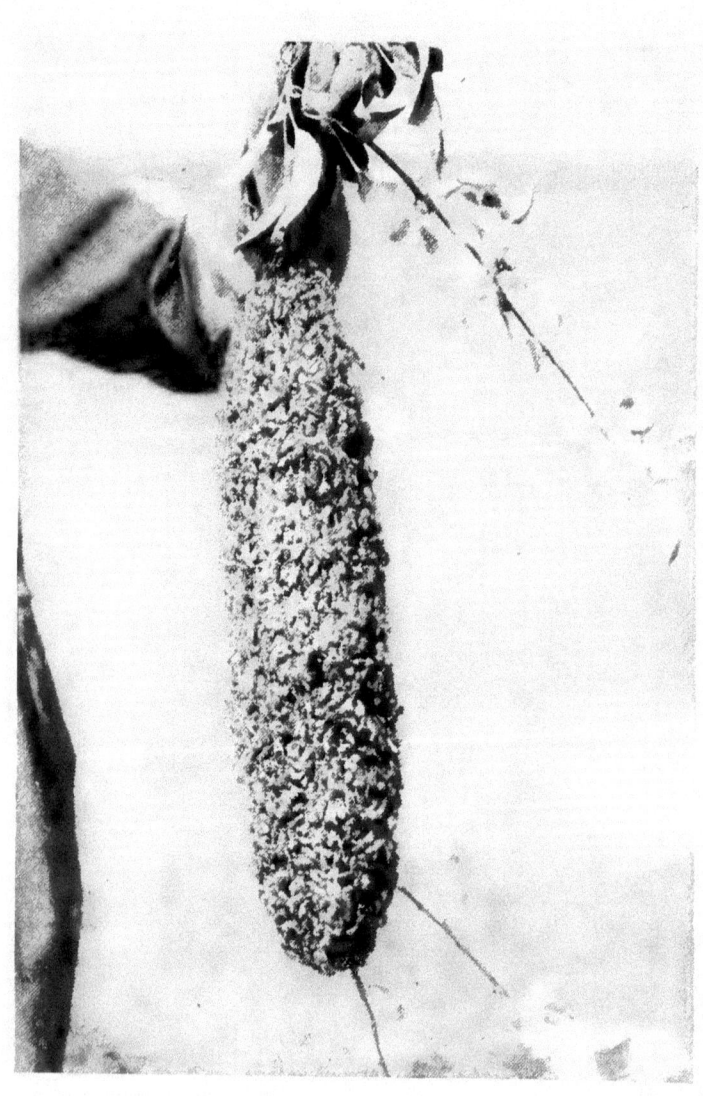

A medium-sized swarm clustered on the branch of an apple tree.

Every Step in Beekeeping

A Book for Amateur and Professional

By

BENJAMIN WALLACE DOUGLASS

AUTHOR OF
Orchard and Garden

ILLUSTRATED FROM PHOTOGRAPHS

INDIANAPOLIS
THE BOBBS-MERRILL COMPANY
PUBLISHERS

SF523
D65

PRESS OF
BRAUNWORTH & CO.
BOOK MANUFACTURERS
BROOKLYN, N. Y.

To

GEORGE S. DEMUTH

THE GENIUS TO WHOM I OWE MY FIRST INTRODUC-
TION INTO THE MYSTERY AND FASCINATION OF
BEEKEEPING, THIS BOOK IS AFFECTIONATELY
DEDICATED.

FOREWORD

The beginner in any field is always more or less at sea. There are always little, fundamental things that he wants to know and too often these same little things are the most difficult to discover.

Often, too, when a man learns enough of any one subject to write a book about it, he forgets or overlooks some of the fundamental things that he learned in his own beginnings and as a result assumes that his readers know as much about his subject as he does himself.

In this little book I have tried to keep in mind my own troubles when I first started to cultivate the festive bee. I then knew the creatures only as an entomologist would know them. To me they were merely a species of insect having certain anatomical characters. I knew their life-history in a perfunctory way,

FOREWORD

but when it came to real knowledge of their habits, I had none.

But for the kindly help of the man to whom this volume is dedicated my flounderings would have been prolonged and I would have accumulated information as I accumulated stings—by a painful process.

A book of this size does not and can not claim to tell all there is to know on the subject of beekeeping in spite of its title. The subject is a big one and it has many sides. The beekeeper must be content to be a student, learning a little each year and with the consolation that men have spent their lives in the study only to leave many unsolved problems behind them.

If the little book serves as an introduction to the facinating business of keeping bees and if it guides the steps of its readers in the right direction it has fulfilled its purpose. The writer has had no other object in view.

B. W. D.

CONTENTS

EVERY STEP IN BEEKEEPING

Every Step in Beekeeping

CHAPTER I

WHY KEEP BEES

"WHAD' yu mean, 'I ought to keep bees?'" said my neighbor Mickey McFarland, in disgust. "Say," he continued, "I get stung often enough in this fruit business without addin' bees to my temptations."

"Yes," I replied, "we all get stung in the fruit business more or less but, paradoxical as it may seem, the more bees we keep in our orchards the less liable we are to be stung—with a short crop."

We argued it out but in the end I convinced the old man that bees were one of the things he most desired in this world and I think perhaps he will remain convinced long enough to go over to Johnnie McCoy's and buy himself a colony or two. That is, if he does not tell "the old woman" about it and let her talk him out of the notion.

As a matter of fact, fruit growers should be more interested in beekeeping than are professional beekeepers. In the long run bees are worth more to the fruit industry than they are to the producers of honey, though many growers are just as skeptical of this fact as was Mickey when I began to talk with him about the advisability of keeping a bee himself. Most of them know that many kinds of flowers are pollenated by insects and that certain fruits are sterile unless fertilized with pollen from a different variety, but they seldom consider that honey-bees are of any very great assistance in this important work.

If every bee in the United States would be killed to-day, the loss in next year's fruit crop would amount to as much or more than the total product of those bees in honey and wax. And yet this benefit to the fruit industry has been considered only as an incidental advantage, one to be accepted as a matter of course but seldom one to be deliberately provided for.

In making my statement as to the effect on a single crop of fruit, perhaps I should be more definite and say that this calculated loss is in a way hypothetical. It would not work out every season in the proportions stated for the reason that the value of bees as fruit pollenizers is not constant year in and year out. For instance, next season might be one of those rare years when the weather conditions are such that pollenization could take place without any very great assistance from insects. In such a case the loss due to the absence of bees would be slight, if noticed at all. But in a bad year for normal pollenation the bees would make all the difference between a full crop and a total failure.

When our trees are given a period of bright, warm, windy weather while they are in bloom we can usually count on a good crop of fruit because under such conditions the pollen blows from tree to tree and is carried long distances by the wind. Also, during such weather there are always hosts of other insects browsing around the flowers and they, in a lesser degree than do bees, carry the fertilizing elements from tree to tree and insure a better crop of fruit.

But if the trees bloom during a period of cold, wet, calm weather we can not expect natural or wind fertilization to be much of a success. Under such conditions the pollen is rendered heavy by the excess of moisture, it does not blow from tree to tree because of its weight, and the wild bugs do not stir out of the sheltered places in which they have taken refuge. But it is in just such seasons that the honey-bees get in their good work. While it is true that they will not fly when the weather is cold and rainy, they will promptly emerge and get to work if the sun shines ever so little. There may be but one hour of sunshine in a whole day, but the honey-bees will be out and clawing their way over the apple blooms while the wild insects are still turning over in their minds whether to go to work to-day or wait until the whistle blows to-morrow.

In some seasons, too, the native insects will not appear in sufficient numbers to be of any great importance as factors in fertilization. Due to previous cold weather they may not have multiplied to the extent that they are numerous enough to be of value. None of them winter over as adults in any considerable numbers, while the honey-bee passes the winter, not as an individual, tucked away under protecting scales of bark or hidden in hollow logs, but in vast colonies, strong in numbers and eager to make the most of the first flowers that appear in the spring.

Honey-bees, too, will vary in the enthusiasm with which they attack their first work in spring. The black German bees are inclined to be of a rather morbid disposition, chary about getting their feet wet in spring and much preferring to sit around the fireplace at home and talk it over rather than hustle out to see what is doing in the orchard. The various "hybrids," crosses of the blacks with better sorts, usually inherit the tavern-haunting tendencies of their unworthy ancestors—not to mention the fact that they carry a dirty dagger in their tails.

The yellow Italian bees, however, seem to have a much more cheerful disposition. They seem eager for the sun and if it shows its face for but a few minutes, and sometimes if it does not show at all, these southern bees will be out hustling for whatever they can find. This is written in January and from my window I can see two "pet" colonies that I have near the house for the purposes of study and ob-

servation. One of them is a colony of
beautiful Italians, pedigreed, regular
royal family stuff you know, and the other
is a colony of Germans that never even
took out their first papers. It is a warm
day for this season, the thermometer is
around forty degrees and there is a rather
pale sun shining. Not a day that one
would think bees would be tempted to fly
much, and yet those Italians are out
skirmishing around. I don't know what
they expect to find at this season but they
are looking for it anyhow. The blacks
are to all appearances dead, though if you
will drum on the hive they will sally forth
as though for world conquest. Consequently, for the fruit grower, the Italian
bee has it all over the dark bees of other
races, simply on account of this tendency
to fly at the least opportunity. Other advantages of both races will be mentioned
later.

The apple, pear, plum, peach, all small
fruits except strawberries, all members

of the legume group, melons, cucumbers, etc., are greatly benefited through the work of the bees. When cucumbers are grown under glass the grower must use some artificial means of pollenating the flowers or he will have no crop. In some cases this is done by means of a camel's hair brush coated with pollen and applied with great labor and patience to bloom after bloom. In large commercial cucumber houses such work would be so great as to be prohibitive and so the grower uses a colony of bees indoors to do the work for him—and they always do it much better than he could do it by hand and at much less expense.

Recently some investigators have shown that bees are essential to the "set" of a good crop of cranberries. I have not seen a full account of the work which led to this conclusion, but it is one case where I am just a bit inclined to think the matter is overstated. I happen to have seen wild cranberries growing in the swamps

of southern Canada in districts where I
know there were no honey-bees within at
least fifty miles. Of course even in such
a case we could draw no definite conclu-
sions. The crop I mention may have followed a favorable blooming season such
as we often have in our apple sections
and in such a season, as I have mentioned,
the value of the bees would be negligible.

It has long been a matter of argument
as to whether or not bees injure ripe
fruit. I have no hesitancy in saying that
they do, though I know that my statement
will at once cause a lot of beekeepers to
rise with a roar of protest. No one is
more jealous of his position than is the
average beekeeper and no one more
eager to argue a question which implies
that he or his live stock may be in the
wrong. Though I am a beekeeper myself, I sometimes like to stir up the clan
just to hear 'em buzz.

The statement has been made so often
by beekeepers that perfect fruit is never

attacked by bees, that they nearly all believe it. They will even prove it to you
by confining some perfect fruit in a cage
with a lot of bees and call your attention
to the fact that the fruit remains un-
touched. If any of the fruit is punctured
so that the juices are easily obtained, the
confined bees quickly attack it. In general this condition of things prevails in
the orchard. It is not often that bees
will deliberately attack perfect fruit on
the trees, but I am absolutely certain that
they do this sometimes. I have repeatedly seen them set to work on the tip end
of a perfectly sound peach that was fully
ripe and claw and suck at the tender skin
until they had made an opening. This,
however, merely by way of argument; I
realize that from a practical standpoint
bees do not injure fruit on the tree.
There are two reasons, however, why they
do not: In the first place there is usually
enough fruit with injured skin on which
they can work easily; and in the second

Old-fashioned hives of an undesirable type.

An orchard is an ideal location for an apiary because the bees secure the abundant flow of nectar from the bloom, and the fruit is benefited by their work.

place most of our fruit grown in commercial orchards is picked before it reaches a stage when it would be attractive to bees.

But lest the reader think that bees are of value only as an influence in fruit growing let me assure him that they also sometimes produce a surplus of honey which is widely known and justly famous as an article of human food.

When I say they "sometimes produce a surplus" I am adhering strictly to the truth because they often do not produce enough for their own use during the following winter. The reason for such failures will be developed as this book progresses. Suffice it to say, however, that the cause of failure is usually with the man who owns the bees rather than with the bees themselves. I am assured on good authority that there are eight hundred thousand beekeepers in this country and that less than ten per cent. of them are good beekeepers.

The annual honey crop of the United States is said to be valued at about forty million dollars and other good authorities who understand both mathematics and beekeeping have assured me that this crop could be increased to three times its present amount if the beekeepers of the country would only give their bees better attention. Unlike many other side-lines this is a business that the fruit grower can engage in and have a double interest and a double return. He will secure better crops of fruit, and as an incident to his improved orchard conditions he will harvest a crop of honey to help pay the income tax or the deficit on the orchard— as the case may be.

Bees do not require close attention for long periods of time. Unlike poultry, or rabbits, or dairy cattle, or goats, or any of the other breeds of live stock, they do not have to be fed daily, they do not have to be milked, and the eggs are not gathered nor the young ones sold at a certain age.

They are a care-free sort of crop if you
want to consider it that way. I knew a
man once who taught school in a little
Indiana city and kept bees "on the side."
He managed to keep busy and out of mis-
chief during nine months of the year by
teaching school, for which labor he received the sum of nine hundred dollars
per annum. The other three months of
the year he played around a bit with some
two hundred colonies of bees and his average crop netted him something like
twelve hundred dollars. Don't jump to
the conclusion that you, too, can without
experience make one hundred dollars a
month out of two hundred colonies of
bees. You can't do it. I know, because
I tried it ten years ago and lost my "roll"
doing it. Since then I have been content
to keep a few colonies, partly for the good
they do in the orchard, partly for what
honey I can get and to a very considerable extent because the creatures are so
eternally interesting. Whether you de-

cide to keep bees for any one of these reasons be certain that you start with a modest number. One good colony is enough for the beginner. More than five is too many.

A few words back I spoke of the bees from the standpoint of interest. They are the most entertaining thing on a farm —after you begin to understand them. A few "hives" set out back of the barn in a neglected corner and forgotten about from one year to the next are not interesting, but a well-kept colony of bees will provide an unflagging interest to the oldest as well as to the youngest member of the family. Last summer a friend was visiting the orchard. He was a city-bred man not familiar with anything on a modern farm. One day I asked him if he would help me with some photographs that I wanted to take of a certain colony of bees. He consented rather reluctantly I thought and then I remembered that he had probably no conception of what the inside of a beehive looked like and no idea about bees themselves except that they would sting, as he supposed, at the slightest provocation.

I supplied him with a bee-veil and gave him my assurance that he would not be stung. He has considerable respect for my veracity, and against his natural fear of the business end of the insects we went to work. His part of the job consisted in holding up brood frames covered with bees, so that I could bring the camera to bear upon them. I, of course, did all the work of extracting the frames from the hive and returning them to their places after we were through. When the ordeal was over my friend drew a deep breath and said, "Well, if any one had told me that you could handle bees as roughly as we handled them and not be stung to death, I would have said he was a liar by the clock."

As a matter of fact, we had not handled them roughly at all—quite the contrary,

though it no doubt seemed rough to the
novice to have me pull out a frame of
brood and see the bees rolling over one
another getting out of the way. During
that entire season I was busy making pictures of the bees in every possible position and naturally handled them much
more than one would if he were keeping
bees only for the honey to be obtained
from them. During the whole season I
received but one sting—and that was a result of my own carelessness.

The life-history of the bee is so full of
interest that books have been written
about it, delightful books too, some of
which are read by persons who never
kept bees and never expect to. Their
habits are different from any other animals we know, and the activities of the
queen read like a fairy tale out of the leg-
end-laden days when man talked with
the wee folk who danced on the lawns in
the moonlight. Imagine the queen, a living creature capable of producing at will

eggs which will hatch into either worker bees, drones or queens. How does she do it? How does the bee know its own hive when it returns from a flight many miles in extent? Why do bees sting more readily on certain days than on others? Do the drones sting and if not, why not? Why does a hive "swarm"? And how does it swarm? Where does the swarm go? Some of these questions have never been satisfactorily answered by any one. Others we can answer by using our eyes and ears when our own bees are in action, and working out problems for ourselves is one of the real joys of life.

To help you answer some of these questions for yourself, and to suggest others to you, will be one of the objects of this book.

Before I leave this chapter, however, I want to say just a word about where bees may be kept. I mean where they may be kept with a reasonable chance of producing a surplus of honey for their owner.

There are probably few places in this
country where bees, if kept in not too
large numbers, will fail to be profitable.
Some sections may easily be "over-
stocked" but such places are already well
supplied with beekeepers. In certain
parts of northern Indiana I remember
that beekeepers sometimes quarreled
over the "right" to certain locations.
Individual apiaries had as many as six
hundred colonies and the pastures were
no doubt crowded; but in the average
farming districts of the country a few
colonies of bees may easily be maintained
at each farm home with no fear of much
more than touching the surface of the nectar supply which is available but unused.
Even in cities bees have been and are being kept with entire success. I remember
that one year when I lived in the city my
secretary kept a colony of bees in the
"back yard" of her home in a densely
populated part of the residence district.
That colony produced something more

than one hundred pounds of honey that season. It should be remembered, too, that the character of the wild plants in the neighborhood has everything to do with the quality of the honey which is produced. Through the "corn belt" we find great quantities of white clover in the pastures. This produces the finest honey known, but unfortunately the introduction of "sweet clover" has resulted in a decided lowering of the quality of the product throughout this great territory. The two clovers bloom at about the same time and a mixture results which is never as good as the "straight white clover." From the North, around the Great Lakes, is gathered wild raspberry honey, thought by many to equal that from the white clover. From the West comes alfalfa honey in quantities. Mesquite contributes its share, while the eucalyptus trees of California are the source of tremendous amounts of a dark honey resembling cough medicine in taste but said to be relished by many who learn to like it. Hosts of other plants furnish nectar from which the bees produce their delectable product. Some of these plants, such as the tulip-trees, furnish the material for strong dark honey that is hardly fit for human consumption. Where such plants exist in large numbers it may be found that the resulting "crop" is hardly worth the candle, but one can find out these things only by experience in one's own neighborhood. I am frank in saying that my own bees seldom produce any quantity of first-class honey, but they occasionally surprise me with a present of a nice white comb containing what is apparently pure white clover product. When they fail in that I have the reward of knowing that they have done their part in looking after certain rather fundamental details of the fruit business and then, as I said once before, they are so eternally interesting.

CHAPTER II

EVERY farm boy knows of "bee trees" in the woods and it is a common notion that bees are native to the United States and that they lived in hollow forest trees long before the coming of the white man. This supposition is not correct because the honey-bee was never a native of North America, but was introduced probably by the very early settlers from Europe.

In their original wild state, in the countries where they were native, bees built their nests in hollow trees, in caves in the rock or in any other sheltered places they could find. Probably in warm countries they even built in the open, attaching their combs to the branches of trees. I know of one such case in Florida in late years, where a swarm of bees, escaping from a local beekeeper, built in a low tree and maintained itself for several seasons.

A "hive" as it is known to-day is simply a house or home for a colony of bees. It may be either a simple box or a more complicated structure adapted to the proper handling and management of the beekeeper.

One of the first forms that the bee-hive took was that of the straw "skep," a conical affair formed of coiled layers of straw rope and arranged with an opening near the base for the bees to pass through. Such skeps served to house the colonies in almost perfect condition so far as the physical welfare of the bees was concerned, but they made no provision for the owner to remove the surplus of honey, and of course their construction was such that it was impossible to manipulate the combs or to exercise any control over the activity of the inhabitants.

It is in this sort of package that a nucleus colony is shipped.

A frame of the nucleus as it is received by the prospective beekeeper.

When honey was wanted by the owner he resorted to the simple expedient of burning a bit of sulphur under the hive and afterward shaking out the dead bees. As it was always the heaviest colonies that were selected for this "treatment" the system naturally reversed the law of the survival of the fittest and only the weak and poorly working colonies were carried over the winter from year to year.

The next step in the evolution of the hive was the introduction of the log "gum." This was simply a section of a hollow log, set on end with a wide board nailed over the top to keep out the rain. A hole was bored near the base for an entrance and sometimes provision was made to place a second section on top to collect the surplus product of the active insects. This was a distinct step forward in bee-keeping methods and many beekeepers continued to use the old "gums" until quite recently. They, like the straw skep, provided ideal winter quarters for the bees and it was seldom that a colony so housed died from cold or from lack of stores.

These primitive hives served their purposes for the early settlers, but it was not until the advent of the nail keg and the soap-box that beekeeping really began to look up. These two couriers of civilization served as homes for innumerable colonies of bees and it is no uncommon thing to find them still on the job and getting away with it in remarkably good fashion. Some "progressive" soap-box beekeepers even go so far as to provide what amounts to movable frames in their boxes and on these frames the combs are built. Additional boxes placed on top of the hive serve to hold any surplus the bees may gather and the "soap-boxer" is convinced that his system is about as perfect as it well can be.

All these makeshifts, however, are crude, clumsy and in the long run expensive. While in any of them the bees may find a home of sorts, they can never be handled as we handle bees contained in modern hives. If disease is present in such a colony it is impossible to detect it without wrecking the house of the bees so that in some states the law prohibits the keeping of bees in such "inspection proof" quarters. It would be better for the bee industry if such a law was universal over the entire country. It would permit the identification and cure of certain diseases and the increase in honey production would be very considerable.

The real impetus to beekeeping came, however, when Langstroth invented what is now known as a movable frame hive. Not until then did apiculture really begin to come out of the bush and get its first shave.

To understand the modern hive we should first have an idea of the way in which bees build their combs in their native haunts. Where they inhabit trees, caves or box hives the combs in which

the young are reared and in which honey
is stored are built vertically in parallel
series sufficient to fill the space they occupy. As a rule these combs are in the
form of various sized plates which are
rather carefully and accurately spaced,
one from another. Under certain circumstances, however, and especially around
the sides of the space occupied, the bees
build short lengths of ''brace comb'' and
this may run at right angles to the main
cluster. Not infrequently too, they may
build ''cross combs,'' or the whole affair
may be constructed ''on the bias'' and
the surface of each individual comb may
present a tortuous aspect. In a wild state
the bees will use as much or as little of the
comb as they may need, adding to it in
seasons of plenty and retreating to a
smaller area in winter and during slack
seasons. In other words, there is no confining element in the wild hives; it is an
elastic unit.

This wild condition naturally includes

features that the beekeeper can take advantage of in designing his hives for it must be admitted that "wild" bees frequently live over the winter in better condition than they do in certain forms of manufactured "houses." The beekeeper, too, can avoid certain obvious defects in the natural hive and it is with a view of combining the good points of the natural hive and eliminating its objections that the beekeepers have striven.

After the invention of the movable frame hive many different individuals attempted to work out a satisfactory method of applying the idea to practical work. The result was a tremendous variety of "patent" hives each in its turn heralded as the last word on the subject. I think I have never known any business that has been more fad-ridden than has that of keeping bees, and while it now shows signs of getting down to brass tacks and behaving itself like a gentleman it is by no means over the mania for experimenting.

It loses its head with astonishing speed and regularity on the least provocation. Consult any bee journal for confirmation of this statement and you are almost certain to find a group of correspondents contributing flocks of letters trying to confirm or deny the latest fad. I enjoy those little arguments because the bee-keepers themselves remind me so much of the activity of the hive, perhaps from long association. But I will say that for the most part their arguments are conducted in a gentle manner and they carry no sting—at least they never seem to use it.

Regardless of where or how a modern hive is built it must conform to two fundamental requirements. It must provide a safe home for the colony and it must be so constructed that the beekeeper can, with the least labor, examine each frame of brood as often as he may find necessary. These requirements have largely been fulfilled in several makes of hives

that are now on the market and for the
most part all makes are standardized as
to the size of the frames that are used.
Even here, however, there is just now considerable argument going on as to the
correct size of the frame, *e. g.,* some workers advocating one a trifle larger than
that which has been accepted as standard
for many years. To adopt any new-sized
frame at this time would mean that hundreds of thousands of serviceable frames
now in use all over the country would become mere junk to be replaced with new
material—to the great advantage of the
bee supply manufacturer.

The typical modern hive consists of a
bottom board, hive body (containing the
frames), super (in which the surplus is
stored), and a cover of some sort to act
as a tight roof for the whole. The bottom
board acts as the foundation for the hive
and as it is always near the ground, if not
actually in contact with it, nothing but
cypress lumber should be used in its con-

struction. Such lumber will resist rot for
a much longer time than any other material available.

The bottom board consists of a series of
short lengths of inch material running
crosswise of the main hive and enclosed
with a binding of other inch material to
serve as the resting-place of the hive
walls. The entrance to the hive is arranged not through but under the hive
body wall so that the raised binding
around the bottom board is simply omit-
ted at one end. This provides space for
the bees to enter. At this entrance end,
too, the floor is extended forward a few
inches to provide a little veranda on
which the bees can readily alight before
crawling into the hive. Most bottom
boards are built so as to be reversible and
can be used either side up. When used
with one side up they allow a space of
three-eighths of an inch high for the entrance while with the other side up this
space is increased to seven-eighths of an

inch. The larger opening is provided during hot weather and during a period when the bees are gathering large quantities of nectar from the flowers. At such times the activity within the hive is so great that the bees would pile up at the entrance in their hurry to get in and out unless plenty of room is allowed. The smaller opening is used in winter and in summer if the supply of nectar runs low. It is also used in the case of weak colonies which might be subject to the attack of robber bees from other colonies. Usually, too, a removable cleat is provided by the use of which the bee entrance can be still further contracted during any emergency. The hive body is the real works of the whole apiary. It is here that the bee makes its home and it is absolutely necessary that the beekeeper be able to ascertain at all times just what his insect workers are doing.

The walls of the hive body may vary considerably. They may be of a single thickness of inch boards or they may be double-walled with the space between either left empty or packed with some insulating material. Beekeepers do not agree as to what is the best form of hive body. It has provided food for argument for many years and is by no means settled.

The double-walled hive has been claimed to be the ideal hive because it afforded such perfect protection in winter, but in actual practice it has been found that bees winter no better in such hives than they do in single-walled structures. In fact it is very doubtful whether the construction of the hive has a great deal to do with the success with which bees survive a winter. I have known colonies to live through winters that saw repeated days when the thermometer went to twenty-five below zero, with no greater protection than the walls of a soap-box. Bees in modern hives alongside of the soap-boxes died with great abandon and regularity. This of course is no argument in favor of soap-

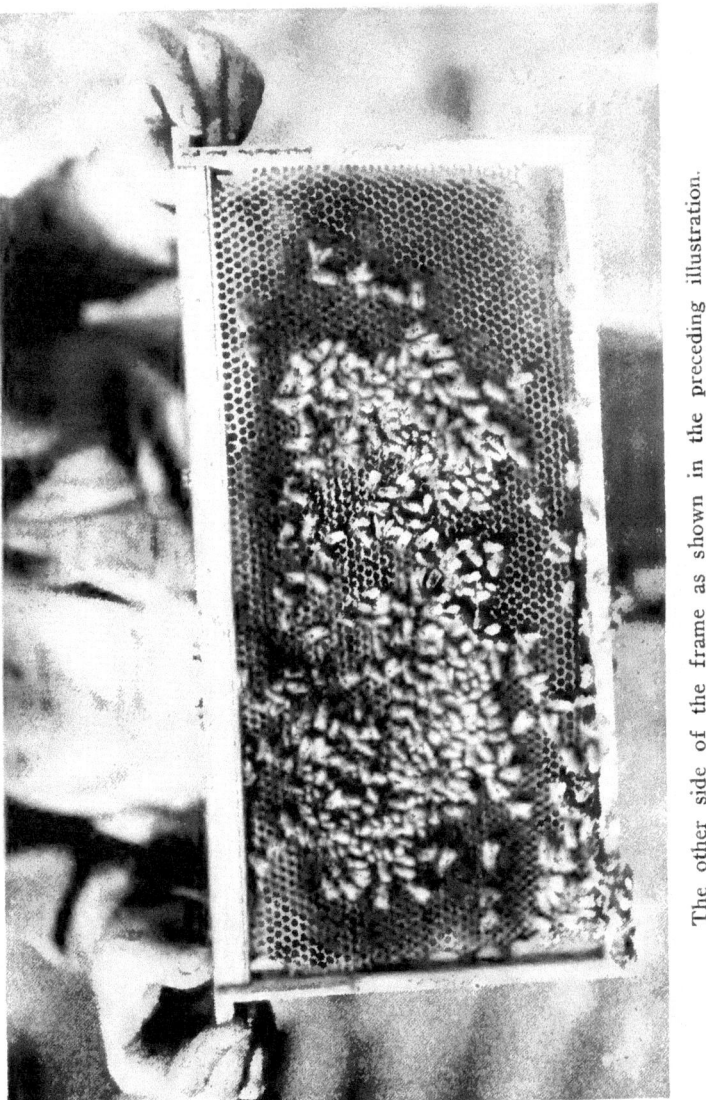

The other side of the frame as shown in the preceding illustration.

A ten frame, double walled hive showing the hive body containing the frames, two of which have been removed. At the left is the inside cover, at the right is the telescope cover.

boxes. It is offered only as an illustration of one of the problems in beekeeping that as yet remains unsolved.

So it is not definitely established that the double-walled hive provides any better home for the bees than does the hive built of a single thickness of lumber.

The walls in any case should be solid and should be joined at the corners with a dove-tail joint to prevent any gaping at these points in case the wood should dry out or warp.

The end walls at the top are cut out so as to provide a place on which the projecting tops of the frames will rest. The frames are the important part of the hive and are now built of a standard size, measuring seventeen and five-eighth inches long by nine and one-eighth inches wide— outside measurements. They are spaced in the hive in such a way that they measure exactly one and three-eighths inches from center to center and this has been recognized as the correct spacing from combs. Bees which build their combs in boxes with nothing to guide them as to the distance between them almost always approximate this distance.

The top bar of the frame should be heavy enough to support the weight of the resulting comb and it should extend beyond the end a sufficient distance to provide ample space to engage the end wall of the hive, which as I have said is cut out so as to allow the top bar to rest below the top surface of the hive wall. All standard frames are now constructed so as to allow room for the bees to work around the ends, top and bottom of the frame, and in this way passage from one frame to another is not interfered with in the least and the activity of the hive can proceed with the least interruption.

In order to get the bees started straight in the building of their combs it is customary to provide a "starter" on which they can begin building. This starter consists of what we call "comb founda-

tion'' and is simply a sheet of pure bees-
wax rolled very thin and stamped or em-
bossed with a pattern corresponding to
the shape of the finished cells.

This foundation is supplied either in
the form of a narrow strip attached the
length of the top bar, or, better still, in
the form of a solid sheet covering the
whole space of the frame. It is held in
place by inserting it in a groove cut for
the purpose in the underside of the top
bar and is further braced with fine tinned
wire extending from end to end of the
frame. This wire is slightly embedded
in the wax foundation and as the bees
build up their comb it becomes a permanent brace which does much to secure the
combs tightly in place. Frames which
are not wired may often be broken in
handling—and that is almost always an
unpleasant experience, especially if the
bees at that particular time happen to be
slightly out of humor about something.

These frames contained in the hive body

constitute what are called the "brood chambers" and they are reserved exclusively for the personal use of the bees in spring and summer for rearing brood, and later for storing food for winter. They are never drawn upon by the beekeeper as a source of honey although similar frames contained in another hive body and placed on top of the brood chamber may be filled entirely with honey as a surplus stock. This honey of course could not be eaten as "comb honey," but would have to be removed from the comb. This arrangement is the source of the extracted honey which is so common on the markets to-day.

The super of the hive is that portion provided for the storing of the surplus honey and may consist either of a hive with standard frames for the production of extracted honey or it may be much shallower and equipped with sections in which the bees can build new comb each time to fill with honey. In the case of extracted honey the same frames are used over and over again and the bees are saved the labor of building new comb each time. Consequently any colony might be expected to return a much larger yield of extracted than of comb honey.

In recent years the tendency has been more and more to the production of the extracted form and as a result comb honey has been scarce and expensive. The high price has resulted in the return of many beekeepers to the production of comb.

The comb honey super is made to fit neatly on top of the hive body, or brood chamber as it is called now. Its interior fittings will vary considerably owing to a rather extensive variety of sizes in which the sections are made. Each section consists of a strip of perfectly white wood, (usually basswood) folded into a rectangular form and joined together by a dovetailed joint. An average super will contain about two dozen of these individual honey boxes.

As in the case of the brood frames, the beekeeper should provide a starter of foundation in each box or section. This is to be neatly stuck to the center of one side of the section in such a way that it will hang down in the exact center of the section. It is best in this case to use full sheets of foundation just as in the brood frames because the full sheets save the bees a great deal of labor and also cause the sections to be filled more evenly and with less danger of building cross-combs. The foundation used in the sections should be very thin as it does not support any great weight and of course no wires are used. The supers are not usually placed on the hive until such time as the condition of the brood chamber indicates that the bees will soon have more honey than they can care for. Whether the super is on the hive or not, the cover or roof is used in the same way as it is interchangeable with either the brood chamber or with the supers.

A good cover should first of all be
water-tight. In the second place it should
not blow off easily. Formerly covers
were made of two pieces of cypress lumber joined in the middle so as to be fairly
tight and with a slight pitch to cause rain
to run off. These wooden covers were a
success provided one always remembered
to lay a few bricks or a cast-iron fly-wheel
or something of that sort on them to hold
them in place. They had a very distress-
ing way of blowing off in the wind—especially at night and in cold weather.
For this reason most beekeepers have
adopted the telescope cover which is built
to slip over the walls of the hive. It is
covered with tin or galvanized sheet-iron
and answers every purpose of a durable
practical roof.

An "inner cover" is almost always
used with these telescope roofs. It consists of a thin light wooden cover to lay
over the top of the hive before the regular cover is put on. Its use provides a

dead air space for insulation against both heat and cold at the top of the hive and it also prevents the bees from sticking down the main cover as they are very much inclined to do.

Perhaps I should at this time say just a few words as to the proper placing of the hives in an apiary. In every case they should be set up off the ground a few inches. Two six-by-six timbers provide an almost ideal rest for a beehive, but any other support which will keep the hive itself from direct contact with the soil will answer the purpose. Also, the back of the hive should be a trifle higher than the front so as to allow any moisture that might collect on the floor to run out at the entrance.

The hives should be placed in such a way that they have some shade during the hottest part of the day. Some beekeepers prefer a location in the edge of a grove for this reason. Others plant grapevines in front of the hive or use some quick growing plant, such as sunflower or castor-beans, to provide the desired shade.

Where a large number of colonies are kept they are better placed in groups of three or four rather than in long unbroken rows. While the bees have a remarkable sense of direction and can find their way home with wonderful accuracy there is no reason for making this job any more difficult than necessary. They have enough to do when a honey flow is at its height without searching for house numbers or asking the corner policeman to point out where they live.

CHAPTER III

BEE BUSINESS

WHEN Solomon admonished the sluggard to go to the ant and observe her ways and be wise, he let his foot slip. Old Sol may have been an artist at picking wives, but he did not know much about ants or he would have understood that their restless activity indicated only a neurotic temperament rather than any wise industry.

If, however, he had held up the honeybee as an example of thrift he would have been entirely within the facts in the case, because the bee not only is industrious but it has a definite end in view to justify all of its labor.

To one looking for the first time at an opened hive of bees much of the activity

42

that is presented to view must seem to be
a sort of lost motion. There appears to
be quite a bit of aimless wandering
around. In a hive that is disturbed by
being opened for inspection, this is doubtless true, but even in a closed hive the bees
are constantly on the move, constantly
doing something for the welfare of the
colony and before we can understand
some of the strange things that we see, we
must know as much as possible about the
normal life of the bee. We must know
what its functions are and how it per-
forms them before we can expect to keep
bees intelligently.

I have mentioned before that there
are many races of bees. The common-
est of these is the German or black bee,
early imported into this country and long
since established in a wild form over most
of our states. Where bees are kept in
crude, old-fashioned hives they are nearly
always of this race either in a pure state
or in the form of "hybrids," crosses with

the Italian bee. These, of course, are not true hybrids but the term has. been so long current among beekeepers that its use is impossible to avoid.

The black bees are often excellent workers, but their disposition to sting on the least provocation makes them unpleasant to handle. This tendency more than any other they transmit to their hybrid cousins and frequently we will find a colony of hybrid bees so cross that it can be handled only with difficulty even by the most expert manipulator.

The more recently introduced Italian bees, on the other hand, are exceedingly gentle and can be handled under certain conditions with very little fear of trouble. They too are excellent workers and offer a combination of an industrious spirit with a mild disposition. For this reason they are fast becoming the standard stock in all well regulated apiaries. Unfortunately, however, there seems to be a tendency among some beekeepers to breed their stock for color and disposition

The hive ready to receive the nucleus.

A model apiary located in the edge of a forest.

rather than for utility. This is a problem
that many live stock breeders have encountered in recent years. Hens in particular have been "bred for feathers"
rather than for eggs and as a result many
"show" specimens are poor egg producers. Poultry men now realize this fact
and beekeepers must soon recognize it.
A few seasons ago I bought a beautiful
Italian queen from a breeder noted for
the excellence of his stock. Her eggs produced the gentlest bees that I ever
handled, but unfortunately they were as
indolent as a Florida "cracker" on a
warm spring day. They were beautiful
yellow fellows with just the nicest dispo-
sitions in the world but they lacked the
pep to get out and hustle for a living.
They finally starved to death one spring,
with a pan of sugar sirup sitting right on
top of their frames.

This case, of course, is an exception,
but it does show that it is possible to
overdo the business of breeding bees for
color and gentleness.

Caucasian, Cyprian and Carniolan bees
have at times been tried out in this country, but none has offered any advantages
over the Italian, and I think that none of
them is looked upon to-day as offering
any particular promise for the future.

Regardless of the race to which our
bees may belong we must know something
of how they act, of what their life-history
consists and how they get along among
themselves.

Each colony of bees contains a varying
number of individuals, ranging from perhaps as low as five thousand or less to
more than fifty thousand, depending upon
the season and upon the general vitality
of the colony. This vast number of individuals is divided into three classes—we
can hardly call them three sexes.

The largest class contains the worker
bees, those who do the real work of the
hive. They go out and rustle up the grub
and then come home and build a place to
store it. They feed the young of all

classes. They defend the hive against invasion, and they invade other hives if they get half a chance. If there is any cleaning up to be done about the place they are the ones who do the dirty work, and if, as sometimes happens, the colony loses its queen and there are no more eggs to carry on the tribe, some of the workers even turn their hands to this delicate task and lay eggs. It must be said in all justice, however, that these eggs never secure the colony any advantage because while they hatch, they produce only drones. In short, the worker is all that its name implies—it does the work of the colony. And in this connection I use the neuter pronoun correctly because the worker is simply an imperfect female. It is capable of laying eggs, which have never been fertilized, and these eggs hatch in due time and produce perfect drones—the males of the colony. However, the worker never (or very seldom) lays eggs as long as there is a queen in the colony.

The workers gather nectar (not honey) from the flowers. This is the sweet liquid secreted by most plants deep in the corolla of the flower and doubtless intended as a bait to lure various insects into the interior of the bloom, thus providing for cross-pollenation. This nectar is not honey, it merely forms the basis for that delectable product; but just how the bee makes the conversion, just what it adds and what it takes away, has never been determined very definitely. It is known, however, that honey contains the sweet elements that are characteristic of the nectar as it comes from the flowers and that it contains, too, a minute amount of formic acid doubtless added by the bee as a preservative. New honey contains enough of the essential oils (perfumes) of the plants from which it was gathered to give it a distinct odor, but after a time this odor vanishes through evaporation. Honey contains less water than does the raw nectar, but students of the subject

have never determined whether this is removed in the honey stomach of the bee during the process of carrying its spoils from
the field to the hive, or whether the loss is
due mostly to natural evaporation after
the honey has been deposited in the cells.
Certain it is though, that a very great
change takes place in the nectar between
the time when it comes from the flower
and when it is sealed in the comb by the
bees. Just what the bee does to it we can
not say but we can be sure it is a busy
creature while doing so. The workers
gather and store pollen in addition to
their stores of honey. To aid them in this
work they are provided by nature with
special hairs on their legs so arranged as
to collect the pollen and carry it most
easily. These hair structures are called
"pollen baskets."

The pollen is used in preparing food
for the first stage in the life of the young
bees as will be seen later.

Another little job for the worker is to

produce the wax out of which the combs are built. It may seem a small task, but from all appearances it is a time-consuming affair. When this part of the work is going on we can open a hive and find clusters of bees festooned from comb to comb in living ladders, each one clinging to its neighbor. I do not know why they assume this form during the process of wax production, but they always do. The wax itself is secreted from tiny plates on the abdomen of the bees and as these plates are formed they are detached by other workers and molded into place as they are needed in the new structure. Such wax is nearly always of an almost snowy whiteness and becomes yellow or brown only after it has been tracked over by millions of the tiny feet of the inhabitants.

The drones are the males of the colony and sometimes they are produced in a wanton abundance that makes one question whether the bees have any particular wisdom or not, for the drone outside of his one function as the male of the tribe is utterly useless. When we consider too, that but one drone of an entire colony can possibly be needed to carry on the race and that hundreds and often thousands of others never do any work at all, it seems surprising that the bees should be so lavish in producing this useless member of their society.

The queen is the only perfect female of the colony. Her activity has been the source of study by some of the greatest scientists and yet it is still only possible to guess at some of the remarkable things which she does. After a young queen is fertilized by a drone she remains fertile during her life and though she may live for four or five years she will continue to lay eggs which are perfectly fertile and which produce sound healthy workers. When we stop to consider, too, that during the height of the spring season a queen will sometimes lay as many as three thousand eggs in a single day we can begin to realize that the job she has ahead of her is a tremendous one.

Another curious thing about the queen is that she can at will lay two kinds of eggs—either fertile ones, which produce workers, or infertile ones that can develop only into drones. These are laid in distinctive sized cells, the ones in which drones are produced being much larger than the ordinary worker cells. The fertile eggs can by proper attention on the part of the bees be developed, not into workers but into new queens. In this way, from one queen we get two types of eggs that may produce three types of individuals.

It is only the queens and the workers that are equipped with those efficient little hypodermic needles known as stings, thus showing that Kipling was absolutely correct in what he had to say about the female of the species—at least so far as the bee is concerned. The queen, however, will very rarely resort to the use of her sting except against a rival queen, keeping the royal weapon for a royal foe. The drones are as harmless as house-flies but they often buzz about one's head in a way that almost makes a person forget that they are pure bluff.

The life-history of these different individuals is not only interesting but it is a very important part of any beekeeper's education.

The egg of the worker is a tiny white, elongated object about the size of the shank of the small letter "i" in the print on this page and about the shape and size in cross section of the dot over the "i." These eggs are deposited in the worker cells in the brood frame and almost immediately are *fed*. Can you imagine any one feeding an egg—and yet that is exactly what happens in the case of the bee. The food is a milky substance provided from the mouths of the adult bees and in this food the tiny egg is immersed for three days. At the end of that period it hatches into a tiny curled, white grub entirely helpless in the bottom of the cell. It is fed constantly by the attending bees until it is six days old when it has increased in size to almost unbelievable proportions. At this time it practically fills the cell in which it now lies at full length and the bees cap it over with a thin, paper-like material. In this sealed cell the immature bee remains for eleven to twelve days longer and at the end of that time it chews a hole in the cap over its cell and emerges a fully developed worker bee.

These young bees can always be recognized for several days after they emerge for they are covered with downy whitish hair. It is their period of adolesence. For a week or ten days after they emerge the workers are engaged entirely with indoor pursuits; they are the freshmen to whom fall all of the cleaning-up tasks. They must take care of the babies and wash the family clothes and see that the floor is swept and look after the innumer-

This large swarm enveloped a young tree and bore it to the ground. Such a swarm contains many thousands of bees.

A swarm enter the hive. Note the newspaper spread in front of the entrance.

able little household duties of the colony. In winter, of course, there are no activities in the incubator department of the hive; the queen takes a rest and at this season the inside jobs must be looked after by older bees. I'll bet they hate it. When the young bee is ten days or so old it is sent out to see what it can find in the way of pollen for the bee babies, the first responsible task with which it is trusted outside of the hive. After it serves its apprenticeship at this job it is allowed to go after the real stuff and brings in its first load of nectar when it is perhaps two weeks old. When it has been engaged in adult activities for a month it is in its prime as a producer of honey.

The life-history of the drone is very similar to that of the worker except, as I have said, it is produced from an infertile egg. In other words, a drone bee never has any father but the fact apparently causes him no worry. In the case of the drones a somewhat longer period elapses between the laying of the eggs and the emergence of the adult—being a male he must be excused for the usual masculine habit of being a bit behind time.

The cells in which the drones are produced are decidedly larger than worker cells and the cappings which the bees build over the larvæ are decidedly elevated. A surface of drone brood looks for all the world as though the cells were full of wind that was bulging them outward almost to the bursting point. If bees are left to build their combs according to their own ideas they will always provide plenty of these large cells, especially around the edges of the frames. This tendency can be in large part avoided by forcing them to build combs only on full sheets of foundation. In this the size of the cell is stamped in the wax in such a way as to induce the bees to build only worker cells. We can always be sure they will find room for a few drone cells around the edges or at the bottom even where such foundation is supplied.

The queen is reared in a peculiar cell of her own. It does not form a part of the regular comb and is not to be found in a hive except during the swarming season. Instead it is built either at the bottom so as to hang downward, or is attached at the side of the comb in a pendant fashion. In shape and size it reminds one slightly of an unshelled peanut stuck on the side of the brood frame.

The egg from which a queen develops is exactly the same kind of an egg that ordinarily would produce worker bees. The difference in the result is due entirely to the sort of food that the egg and larva receive. This food has been termed "royal jelly" and by removing some of it from queen cells that are started and dividing it among a number of artificial cells one may raise as many queens as one likes. This, however, is a mighty particular job and one that the beginner, the amateur or even the average beekeeper of some experience can well leave to the specialist.

I have mentioned the fact that the queens are produced only during the swarming season. The reason for this is that when a colony "swarms" the old queen always leaves with the swarm and some provision must be made for a successor to the absent monarch. The swarm is merely an exodus of the older bees of the colony in search of new quarters. It is the natural way in which the bees increase their numbers and the instinct is so strong in them that the beekeeper must be on the job constantly during the spring of the year when honey supplies are piling up in the storehouse. It is an instinct that is emphasized by the crowded condition of the hive at that season and often many things can be done to discourage the idea. These manipulations, however, will properly come in a later chapter.

There are many things about bees that are not known and probably never will be. One of these is the ability of a bee to recognize a person who is afraid of being stung. That they do this is almost certain for there are some persons who are utterly unable to work with bees on this account. I once had a man working for me in the orchard who was mortally afraid of bees and if one came within reaching distance of him he was certain to be stung. He never attempted to open a hive and I often wondered just how badly he would have been stung if he had attempted that feat. Probably not so much as might be expected for it would have required courage on his part to handle bees and the bees in turn would have known that he was less afraid of them. With proper handling there is not much chance of being stung, but at the same time one must take certain precautions and the beginner certainly should be urged to watch his step. Years ago when I first began to write about bees I resolved that I would never publish a picture showing any one handling bees without a bee-veil on. It is a regular gallery stunt to open a hive of bees without either smoke or veil and many writers disdain the use of a veil when posing for bee pictures. Much harm has been done by this practice for the beginner has been loaded full of a cocky assurance which makes him think bees do not sting under any circumstances, and then there comes a day of reckoning when the bees are feeling peeved about the last election or something similar and the novice gets a dose of stings that dampens his enthusiasm very considerably.

Therefore, never attempt to work around a colony of bees without a good bee-veil over your head. It is best to put the veil on over a wide brimmed straw hat as the hat will hold the netting away from your face. I usually turn up the collar of my shirt and tie the lower ends of the veil under my necktie. In this way I make a bee-tight joint around my neck and have no fear that any crawler may suddenly appear on the tip of my nose.

Some beginners use "bee gloves," heavy canvas affairs so thick that the bees can not easily sting through them. When I first began to handle bees I thought the glove idea was a good one and as I had at least one colony of rather cross individuals I put on a pair of leather gauntlets for protection. The difficulty with them was that my fingers were so stiffened that I frequently jarred a frame of brood when I should have handled it with care. Repeated errors of this sort so aroused the bees that they went on a stinging spree and quickly found the funnel-like entrance to the gauntlets—to my great discomfort for several days afterward. My wife, watching me from what she thought a safe distance, was noticed to return suddenly to the house, walking in a stiff and awkward manner unusual with her.

A good bee smoker is just as essential to a well managed apiary as are the bees themselves. It is a tin contrivance with a small bellows attached, with which one can produce an abundance of smoke at will. The best material to use for fuel in the smoker consists of old cotton rags or pieces of burlap. An old burlap bag rolled rather loosely and cut up into two or three inch sections tied with string makes excellent fuel for this purpose.

I do not know just what effect the smoke has on the bees but it certainly has a quieting influence on them and the crossest colony can often be handled with impunity after being smoked.

In opening a hive it is always well to blow a little smoke in at the entrance, not much, just enough to announce to the bees that something is about to be pulled off. Then, pry up the cover about a quarter of an inch and shoot somewhat more smoke into the hive. If the colony is a gentle one but little smoke will be required, but if it is cross or if the weather is cool or damp more smoke should be used.

Remove the cover firmly but gently so

as not to jar the hive more than is necessary and blow a little smoke over the tops of the frame. At this the bees will usually buzz actively but in a not unfriendly way. By experience the beekeeper will learn something of the various notes of the bee and can detect a decided difference between the noise they make when they are angry and when they are not.

Watch the hive closely while working with the brood chamber and if you notice the bees collect in the spaces between the top bars in solid rows and each bee with his head pointed outward you may know that it is time to give them a little more smoke. If they begin to fly about you with a high keyed note you should give them still more smoke but never smoke a colony of bees that is quiet and that appears to be going ahead with its work in a peaceful way. Occasionally a single individual will become so aroused that nothing short of sudden death will satisfy it. Such a bee will fly around your head and buzz angrily in front of your face intent upon stinging you. There is only one thing to do to such a hot-head—send it to the morgue.

Above all, in handling bees try to avoid pinching or crushing them when you remove frames of brood from the hive. Nothing you can do will anger them more quickly. You may drum on the hive with a hammer; you may shake or brush the bees from the comb on to the ground; you may jar the combs suddenly so as to remove the bees and they will often remain docile; but if you accidentally pinch one it will be after you with a rush nine times out of ten. Frequently it takes only a leader to start a mob and, believe me, when the bees start out on a real Bolshevistic orgy they can out-Lenine Lenine himself.

CHAPTER IV

HOW TO GET THE BEES

GETTING a start with bees is somewhat
different from securing any other kind of
live stock that I know about. These little creatures, in spite of all the care given
them, are still very much in their wild
state as far as their life habits are concerned. We have learned to handle them
so that they remain within reasonable
bounds, but when the swarming instinct
comes over them they revert to the habits
of their wild ancestors of thousands of
years ago. By swarming the bees increase their numbers. It is the natural
way in which a crowded colony makes
room within its walls and is a provision
of nature to make sure that the race will
not die out. It is little wonder then that

we have been able to do only a trifle to counteract this instinct for it is one of the strongest forces in existence not only among the bees but in all other animals and plant life. To nature, the reproduction of the species, the continuance of the race is the only thing that matters. It is her one big question and she always answers it in her own way. Therefore when a colony throws off a swarm it is reverting to its wild state and unless the beekeeper is present to catch the swarm when it issues from the hive it will in all probability wander off and take up its abode in some hollow tree, perhaps miles away from the parent colony.

It is in this way that we happen to have "wild bees" in our woods. They are in every case bees that originally escaped from some beekeeper. Or we might argue that all bees are really wild and that our efforts at domesticating them have been more or less of a failure since they return to their natural condition at

Transferring bees from old hives into new ones. In this case many of the frames were simply lifted out of the old worn boxes and placed in the new bodies.

A brood frame showing three queen cells.

the first opportunity. They not only return but are able to maintain themselves in a wild state—something that could hardly be said of any of our domesticated animals with which we stock our farms.

The new beekeeper therefore has at least one source of stock upon which to draw that is not open to other "stock raisers." Even in Jersey one can not go out and catch himself a Jersey cow anytime he wants it and if I am correctly informed, Shanghai roosters are as carefully cooped in Shanghai as they are in Boston. Wild bees can be obtained from the woods in many parts of the United States and in the more remote portions they are sometimes very common. All one has to do is to go out and locate a bee-tree and remove the bees. Sounds simple, all right, but there are strings to the scheme just as there are to all other easy-money propositions. In the first place the beekeeper must locate the bee-tree, which is not the easiest thing in the world for a novice to do. The method, however, is so interesting that I believe every one would be glad to know how it is done even though one never expects to try it one's self.

The bee-hunter prepares a small amount of diluted honey for bait and takes with him a small box three or four inches square, with no bottom and with the top made of glass. Fastened to the side of this box is a small dished shelf into which a small quantity of the diluted honey is poured.

After a place is found where the bees are feeding on flowers, a few individuals are caught by placing the box over them. They crawl up the sides, find the honey and begin to fill up on the rich discovery they have made. After several bees have been caught and before any of them are sufficiently laden to cause them to want to return to their hive, the box is placed on some elevated object such as a stump and the glass lid carefully withdrawn.

The hunter then quietly slips back a

few feet and crouches down so as not to
be in the line of flight of the bees if they
should chance to come in his direction.
Also by taking a position lower than the
hunting box he can more readily observe
the bees as they fly for they are outlined
against the sky.

As soon as a bee has taken on as large
a load of the diluted honey as it can com-
fortably carry it will rise from the box,
make a few circles and then suddenly dart
off in a direct line toward its hive. This
line of flight is so straight that by following it with his eye the hunter can tell
the exact direction of the home of that
particular bee. While he has the direction he, of course, can not know the distance and to obtain some idea of this he
must now change his position to a sufficient extent that he can get a "cross line"
on the first flight. The point where the
two lines come together is probably the
position of the hive or tree in which the
bees live. This method of locating bee-

trees is a very old one and was used years ago.   You will find it mentioned in Cooper's *The Beehunter.*

After lining the bees as above described, the hunter must locate the tree by going into the woods and carefully watching every likely trunk or branch. It requires keen eyesight to do this and the average inexperienced person might pass by a bee-tree a dozen times and never notice it. By keeping the trees between you and the sun it is often possible to see the bees working in and out of a knot hole high above the ground, while with the light at your back they might be almost invisible.

After the tree is located the next job is to get the bees into a modern hive and this is quite another story.

The colony may have constructed its nest close to the ground where it can easily be reached by the hunter or it may be located in a hollow branch seventy-five feet or more above the earth. If the tree is on another person's land it will be necessary to obtain the owner's permission before such a tree may be cut, but I have never had any trouble in obtaining such permission. Usually the owner is willing to have such trees cut because they are as a rule hollow and worthless as timber and the chance of obtaining a few pounds of honey in exchange will tempt the average owner to part with the tree.

If the nest is high above the ground and in a thin shell of wood, it will quite often be badly wrecked when the tree falls. In some cases the wreck is so complete that practically nothing can be done toward saving the remains, but if the trunk or branch holds together it is an easy matter to chop a hole in the wood large enough to remove the combs one at a time. In such cases one should select those combs that are the most regular in shape and that contain plenty of brood or stores. These combs are taken out of the tree and cut into sections as nearly as possible the exact size of the brood frames.

After carefully fitting the combs into the frames the hunter should take ordinary cotton string and wrap it round and round the frame in such a way as to hold the comb firmly in place. As soon as a frame is filled it should be placed in the hive body and the whole job should be done as quickly as possible without hurrying. It never pays to "hurry" any bee work. The little fellows quickly learn that the operator is hurrying and nervous and they in turn become even more nervous than the worker and thus a "vicious circle" is formed that sometimes results in the beekeeper suffering a very sudden desire to go away from that particular place and seek other employment.

While this work is being done the bees will, of course, be flying around and crawling over the combs in a more or less excited fashion, but it is rather remarkable that if they are carefully handled they show comparatively little inclination to sting under such circumstances.

When bees are bought in old box hives
or in log "gums" the same process of
transferring the combs to the frames of
the modern hive must be gone through
with. I remember the first time I ever
saw this work done. I had bought my first
bees from an old-fashioned farmer, pay-
ing him, I think, two dollars a colony for
them. Most of them were in sections of
poplar logs about thirty inches high and
fifteen to eighteen inches in diameter.
I had no idea how to get the little workers transferred to the new hives I had
bought, but a beekeeping friend initiated
me in the process. He first puffed a good
deal of smoke under the lower end of the
log and then asked for an ax. About that
time I began to think that I would much
prefer to keep chickens or Belgian hares
or pigeons or some other gentle beasts.
The idea of opening a hive of bees with
an ax seemed to me like flying in the face
of Providence, but that is exactly what
George Demuth did on that occasion.

With one swift stroke of the ax the board cover, nailed to the top of the "gum," flew off and almost immediately the bees were greeted with a rich puff of acrid smoke which caused them to change their minds about pulling off some reprisal stunts they had all cooked up. Almost before they had become quiet after the removal of the cover, George sailed into the log itself and with blow after blow finally succeeded in splitting the tough old tree lengthwise. I, the novice, stood by wondering why on earth we were not both stung to death.

As a matter of fact the bees behaved themselves with rare self-control considering that to all appearances the end of the world had arrived for them. I learned, however, that when bees are roughly attacked in this way with their whole communal life threatened, they invariably make an attempt to save what they can out of the wreck. To this end every bee sets to work to fill itself as full

of honey as it can possibly get and a "full"
bee either can not sting readily or the desire to sting is less active at such times.
At any rate that first experience at transferring bees from an old gum hive did
not cost either of us a single sting. That
first experience too taught me that the
strings used to tie the brood combs in
place would later be removed by the bees.
They quickly cemented the combs to the
frame with new wax and then they pro-
ceeded to cut the strings into sections and
remove them from the hive. A few days
after the transferring began we found
bits of string in front of the entrance of
the new house. Some pieces, of course,
became stuck fast to the frames and those
we removed at the first opportunity and
saved the bees the labor of pulling them
out of the hive. A bee seems to dislike
anything of a worldly nature and for this
reason the strings were gotten rid of as
quickly as possible.

Securing wild bees from a bee-tree or

buying boxes of bees in crude frames and transferring them constitute the two most interesting and at the same time cheap·est ways of obtaining a start with bees. Neither way is the better way for several reasons. In the first place bees obtained in this way are liable to have brood diseases present in the combs and it is impossible to detect these troubles until after the tree or box has been broken open. It is a case of buying a "pig in a poke," sort of a grab-bag affair in which the buyer never knows what he is going to get until after the deal is closed. In general, however, if a colony of bees is bought in a box hive it is reasonably correct to assume that it is a healthy colony if the bees are abundant and if there are plenty of stores. These two factors can be determined by watching the flight from the hive and by lifting or weighing it. If but few bees are flying during a period when many should be afield and if the box is light in weight we can be almost sure that something is wrong.

A "close up" of three queen cells built at the bottom edge of the frame. This is a favorite location.

Opening a hive, No. 1. Smoking the entrance.

Even with the best of colonies the combs
may be built in such a crooked fashion
that little can be done with them when it
comes to fitting them into the new brood
frames. And if the combs are perfectly
straight and we secure a good fit in the
frames we still do not have an ideal arrangement, for these old combs are never
fastened as solidly to the frame as they
should be. In hot weather or when they
are loaded with winter stores they not infrequently break down of their own weight
for they lack both a good attachment to
the top bar and the brace wires which
should always be present in a properly
planned brood frame. The best way for
the beginner to start with bees is to buy
them in modern frame hives from some
reliable apiary. In that way the trouble
and annoyance of transferring from the
old boxes to the new hives will be avoided
and after having learned something of the
handling of the insects the new owner can
later increase his numbers by the purchase
of stock in the crude hives.

Even when buying bees in a modern hive, certain precautions should be taken to see that one does not unwittingly get more than he had bargained for. I have indicated that it is fairly easy to tell when the colonies are strong, but it requires an expert to determine whether disease exists and whether the colony has a good queen. Both of these items should be investigated before the purchase is made. Even if the queen herself is not observed her presence can usually be taken for granted if we find freshly laid eggs in the cells of the brood frames. These are easily recognized at the bottom of the cells, usually in sections of the comb adjacent to and around the edges of the clusters of cells containing the larvæ in various stages of development. The brood diseases, however, are difficult for the novice to detect, and before buying he should always call in some one capable of recognizing these disorders. In another chapter I shall go into considerable de-

tail regarding bee diseases, but at this
time it is sufficient to warn the prospective beekeepers that such diseases exist
and are widely distributed.

If one buys his stock of bees in good
hives he will naturally pay much more
for them and many beginners have indicated a desire to purchase good stock and
at the same time not have to lay out so
much money right at the start. To supply this demand many beekeepers now
make a business of selling bees either by
the pound or by the "frame." One or
two frames of brood, partly covered
with adult bees and accompanied by
a laying queen is called a "nucleus."
This furnishes the easiest and cheapest
method for the beginner to get his start.
Having bought a hive and become familiar with its various parts he should purchase his nucleus and install it in the prepared home.

Distance is but little object when it
comes to buying bees in this way for they

can be shipped hundreds of miles in perfect safety and one can obtain stock from the best apiculturists of the country and feel certain that the stock he receives will be healthy and of the race which he desires. I have seen bees so shipped several hundreds of miles only to arrive in perfect condition and with practically no dead bees in the bottom of the cage. They are shipped in strongly built crates holding one, two or three frames and with a wire screen at the top and bottom to provide ventilation.

When these shipping cases arrive at their destination they should be carefully opened and the frames inserted in the permanent hive in place of empty frames which are removed to make room. In nearly all cases, bees so obtained will be very short of stores. They will require careful watching to make sure that they do not starve to death and they should be protected against robbers from other hives. To do this, the entrance to their hive should be contracted to as small an opening as will allow only one or two bees to pass out at a time. Under such conditions even a very weak colony can defend itself against intruders from other neighborhoods—and bees will sometimes fly considerable distance to steal from weak estates. There is no self-determination among the small nations of the bee world. It's a case of the survival of the fittest every time.

The nucleus should be considered simply as a very weak colony of bees for that is what it really is. It must be given protection and food in order that it may increase in numbers as rapidly as possible. It is best to place the frames of the nucleus in the middle of the hive body with an equal number of frames fitted with sheets of foundation on each side of the bees. In this way they will be encouraged to draw out their combs equally on both sides of the brood nest.

They should be fed all the sugar sirup

that they will eat and this can readily be furnished to them from a small tin pan placed inside an empty super and covered with the regular hive cover. Regular devices are furnished by dealers in beekeeping supplies for the purpose of feeding bees, but I have had so much better results from the use of the open pan that I do not care to bother with the patent feeders.

My plan is to remove the interior fittings from the super and place the empty body on the hive in the regular way—just as though it was to be used for honey production, except, of course, that it contains no sections. Inside this super and directly on top of the frames I place a small tin pan holding about a quart. Fairly thick sugar sirup is poured into this pan and a handful of twigs, grass or chips is thrown on the surface of the sweet. This débris enables the bees to get at a larger surface of the food and if they should happen to fall in, as they often do, it will enable them to crawl out more easily. If this precaution is not taken you are apt to find your tin pan only a trap for your bees. One should be careful about feeding honey to bees unless he knows that it comes from a healthy apiary. Honey produced by diseased bees can infect a healthy colony to which it is fed. For that reason sugar is preferred to honey for artificial feeding. However, it is not advisable to feed the new colony too much lest it get the idea that it will always have an abundance of food ready at hand and will fail to set to work and build up its numbers. Bees are like humans in many regards. They do not stand prosperity any better than we do and when things are coming their way with a rush they are inclined to loaf on the job a bit and I imagine would take to silk shirts and sport model motors just as readily as people would—if they had a chance.

Therefore it is a good idea to give them just what food they need for their housekeeping and to encourage them to expand their field of activities, but not enough to make them feel that the world owes them a living whether they work for it or not.

With proper care and feeding a nucleus with a good queen should build itself up to a full-sized colony of eight or ten frames before fall and should have enough surplus honey stored away to carry it safely over the winter. Often, however, such a start will increase in strength slowly and usually will require some additional feeding to provide enough stores for safe wintering.

On rare occasions during "good years" a nucleus obtained early in spring may not only store enough for its own use but enough to give the owner a small surplus. This, however, is not the rule and the beginner will do well to be content with a slow start but a sure one rather than to hope for great things and be disappointed.

CHAPTER V

In an earlier chapter I made the statement that there are eight hundred thousand beekeepers in the United States. The term "beekeeper" is a correct descriptive name for the majority of men who engage in this industry because that is what they literally do—they "keep" bees, but they do comparatively little in the way of managing or controlling them. In fact the business of honey production is to-day in but little better condition than was the fruit industry twenty-five years ago. If a crop is harvested it is more than likely due to the abundance of Providence than to foresight on the part of the owner of the bees.

There are, of course, exceptions to this

statement for in many parts of the country expert apiculturists manage their bees with consummate skill and their crops are less dependent upon the seasonal chances than are those of their lesser brothers to whom the bee is but an incident in other work. When I say that the majority of beekeepers do not manage their bees but allow them to manage themselves, I speak of the vast army of "little" beekeepers who have from one to a dozen colonies.

All of these lesser lights could vastly increase their honey production by very little extra care and attention. The ultimate end of beekeeping is the production of honey and if this end is to be gained, the owner of the bees must so manage his insect workers that they are in a position to accomplish the greatest amount of labor in the least time.

While bees gather some nectar from flowers during the entire summer it is only at certain periods that they secure

Opening a hive, No. 2. Lifting the lid and blowing a little smoke over the brood frames.

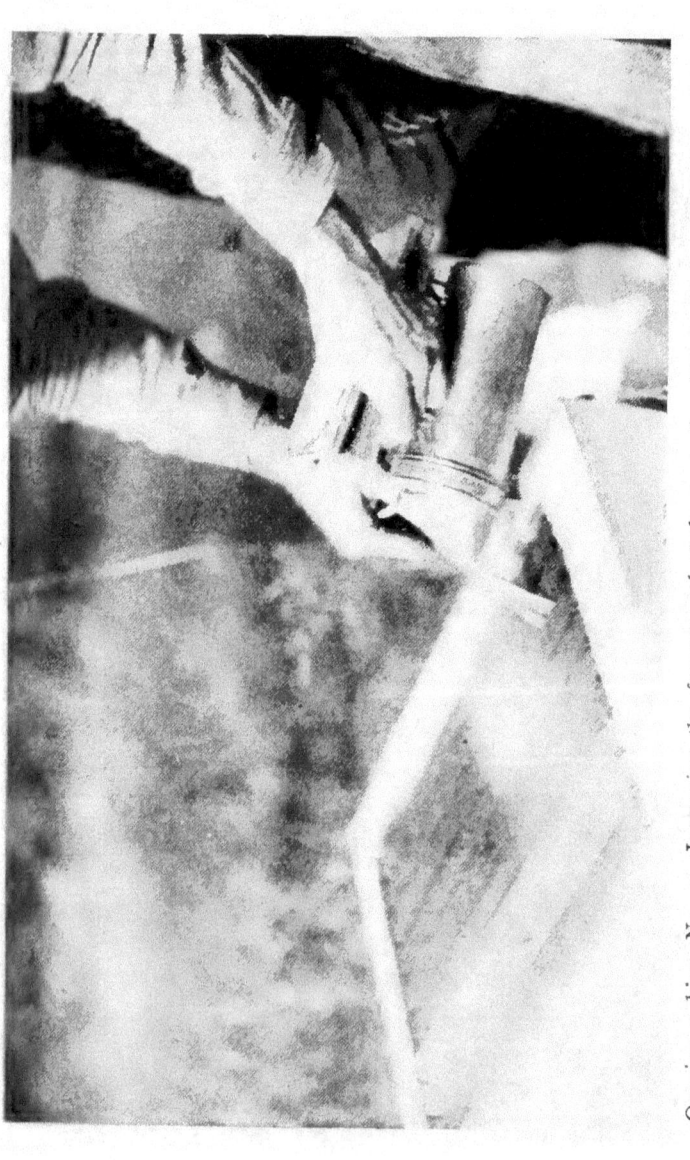

Opening a hive, No. 3. Loosening the frames and at the same time blowing some smoke over them.

more than enough for the daily needs of the hive. These periods are called "honey flows" and their occurrence varies with the season and with the situation of the apiary—both as to its situation in a given neighborhood and its sectional position with regard to the rest of the country.

In the Central States we have our greatest yield of honey from white clover. In parts of Michigan wild raspberry produces the bulk of the crop. In New York, buckwheat furnishes honey late in summer. In a few places basswood yields almost at the same time as white clover. In the West, sage, mesquite and other plants produce their crop. All of these have their period of maximum production and these periods (honey flows) sometimes may last for only a few weeks or even for a few days.

If the beekeeper does not have his bees ready for the harvest when it comes, it will remain in the field and soon enter the same class as water that has passed below the mill. Consequently the individual beekeeper must know something about the honey producing plants of his neighborhood and must adjust his system of management to suit his own conditions. I am saying this partly for my own protection because no system of hive management could be applied to all parts of the country with success. I can only outline a system that works in the white clover belt and I will try to adapt it to other conditions where it seems possible to do so.

Regardless of the location of the apiary or of the sort of honey that is to be produced, the first thing the beginner should learn is that a season's management of an apiary begins in the summer of the previous year. The bees which live over winter are all hatched from eggs laid in August and September and unless a colony has a good queen and plenty of surplus stores it will not rear enough young bees to winter as a strong colony. Given the proper number of bees the colony must have twenty-five or thirty pounds of surplus honey for winter feed and to start brood rearing in the spring.

A strong colony with plenty of stores and housed in a proper hive will usually winter in good condition—although I will have more to say about this in a later chapter.

In very early spring the colonies must be examined to determine whether or not they have sufficient stores to last them until the first nectar-producing flowers open. It might be supposed that a colony that was sufficiently provided in the fall would not require this early examination, but such is not the case because in some seasons bees will consume much more of their reserve food than they will at other times. A warm open winter always results in greater hive activity and as a result more honey is used. Such a winter, too, induces very early brood rearing and as a result the bees will sometimes have a host of young ones to feed in early February and thus add a still greater drain upon their resources. It is not necessary to remove the covers from a colony to determine if they are short of stores. After a very little practice the beekeeper can tell by lifting the hive whether or not it contains any surplus honey. If it is found to be short of stores then provision must be made to feed the bees with sugar sirup prepared by boiling cane sugar and water in the proportion of two parts by weight of sugar with one of water. This sirup can be placed in a shallow pan into which a handful of leaves, grass or sticks has been thrown. The "trash" will afford a chance for the bees to crawl out without drowning in case their enthusiasm gets the better of their equilibrium. These pans are to be placed directly over the brood frames in a super from which the interior parts have been removed. The lid or cover should be returned to its place with care because it is

highly desirable to conserve the warmth
of the hive at this season. After this first
examination and feeding, the bees should
not be further disturbed until warm open
weather arrives when each hive should be
given a thorough inspection from top to
bottom. At this time, as in the first case,
one must look to the condition of the larder
and if it is found to be reduced to the condition of Mother Hubbard's cupboard the
feeding performance must be repeated.

At this time, too, the hive should contain a large quantity of young brood and
the presence of a laying queen should be
attested by the presence of an abundance
of fresh eggs. (April eggs are just as
welcome to the beekeeper as they are on
the breakfast table). If no eggs or brood
are found the colony is certainly queenless
and should be combined with a queenright
(normal) colony. This can readily be
done by simply placing the queenless hive
over the hive body of the normal one. To
prevent any chances of a scrap between

the individuals of the separate hives, it may be well to interpose a thin sheet of paper between the two. A small hole can be made in this and the bees will gradually mingle without going to law over the line fence.

At this second examination, too, the wings of the queen should be clipped—if this little operation has not previously been attended to. Clipping the wings of the queen may sound like a fearsome job to the novice. I know of one novice who was quite terrified when he first attempted it, but he quickly learned that it was not so dangerous or so difficult as it sounded. In fact the only danger is to the poor queen herself for she will not sting even if handled quite roughly. The beginner is quite likely to crush her in his clumsy fingers when he tries to pick her from a brood frame and to prevent any such misfortune I suggest that he practise picking up drones and clipping the wings *on one side only*. After he has learned to handle drones he may further his education by learning to pick up workers by the wings only without being stung. When he can do this stunt he may safely proceed to catch the queen and clip her wings with little fear that he will cause her serious injury.

The object in removing the wings from one side of the queen is exactly the same that we have in mind when we clip the wings of some active old hen that persists in flying over the chicken yard fence. A "clipped" queen will not be able to lead out a swarm but will instead hop around in front of the hive until she is tired and then go back and behave herself—for a little while.

Clipping the wings of the queen does not prevent swarming, but it enables us to have far better control of such swarms as do issue from the hives.

Another thing to watch for at this spring inspection of the hives is disease. There

are several bee diseases that must be
watched for but these too will be considered in a separate chapter. After this
first warm weather examination the colonies must be inspected at frequent inter-
vals so that the owner may be sure that
everything is progressing as it should.
The object of hive management is to have
the colonies at their maximum population at the time of the honey flow and the
owner must adjust the hive activity in
such a way that this result is brought
about.

In localities where the chief honey crop
is harvested early in summer everything
must be done to discourage swarming,
while in sections where the main crop
comes in the fall, as it does in most sections, the colonies may be allowed to increase by swarming as the owner will then
have time to build up both the parent colony and the swarm to full strength before
the crop appears.

There has been much written on swarm

control and many hives have been designed and some of them patented, which were supposed to prevent the bees from accomplishing their desires in this respect. We must remember, however, that it is by swarming that bees increase their numbers and establish new bee communities. It is the instinct for the preservation of the race and when we confront natural laws of this sort we must make allowances. Sometimes we may become impatient with the bees for their apparent insane desire to swarm, but at such times we do not want to forget that they are backed in their desires by thousands of generations of other bees that never knew any better than to go ahead and preserve their kind as their Creator designed them to do.

The modern beekeeper takes these things into consideration and while he does all he can to prevent swarming at the same time he makes the best of it when a swarm issues in spite of his precautions. As a rule bees will not swarm unless they are crowded for room. They will seldom swarm unless they have made provision for a new queen to supplant the old one which always leads out the swarm. These two facts give us a foundation on which can be built some sort of system of swarm prevention.

We can nearly always give them more room either by removing frames of brood which we place in weaker colonies or by supplying them with an additional hive body placed directly on top of the original one. Recently, too, there has been an inclination to adopt a hive containing ten frames instead of the one which was long standard and contained only eight frames. Some beekeepers have maintained for years that the ten frame hive tended to reduce swarming simply because it afforded the bees more room in which to work.

By weekly examination we can detect queen cells which are being built and remove them. If this is done during the season of greatest brood rearing it often results in swarm prevention, but if a single cell is overlooked we may be quite certain the bees will swarm. Often too, they will get the "swarming fever" in their veins and even though the queen cells are removed they go ahead and swarm anyhow. They are almost certain to do this if they once get a good start on a queen cell so we must nip these royal cradles "in the bud"—to change the figure of speech.

Sometimes, too, a colony will apparently go mad with the desire to swarm in spite of all that can be done. I have seen them insist upon swarming, not once but repeatedly, until the parent colony was almost destitute of bees to care for the young brood. One year I recall, the bees in our section had a very peculiar habit of throwing off what we called "crazy" swarms. They would swarm out with every appearance of intending to leave the neighborhood and after half an hour

spent in the air would return to the hive
and go ahead with their work. This joy-
riding continued day after day for a week
or more and I never did understand just
what it was all about. At the time I know
I was a very much worried beekeeper.

When a normal swarm issues from a
hive accompanied by an unclipped queen,
it will settle on any convenient object in
the neighborhood before going to its new
quarters—wherever they may be. The
cluster of bees may form on a low shrub
a few feet from the hive or it may decide
to go to the top of one of the tallest trees
in the neighborhood. It is to prevent any
such performance that we clip the queen's
wings so that she can not fly. The
result of this precaution is to have the
swarm come to a cluster about the queen
close to the hive or in some cases to return
to the hive. In either event the beekeeper
must do something to satisfy the instinct
of the bees. Usually this means providing
them with more room.

Opening a hive, No. 4. Lifting out the frames gently. The smoker is held between the knees ready for action.

Opening a hive, No. 5. After they are once disturbed most colonies can be handled with but little smoke but that is no reason why the smoker should be dispensed with.

While the swarm is out of the parent hive it should be set to one side and its place taken by a new hive body containing empty brood frames. The supers which were on the parent hive are now placed on the substitute hive body. If the bees return of their own accord they realize that there has been a change in the old house and that things look pretty empty around the diggings which they so recently left in a crowded condition. As a result they settle down to work and forget about swarming for the rest of the season. If they do not return of their own accord but settle on some low bush to which the queen has crawled, the cluster should be removed by cutting off the branch to which it clings and be carried over to the entrance of the new hive. A newspaper spread on the ground in front of the hive will be of assistance at this point. The branch is then given a vigorous downward jerk and the mass of bees thrown on to the paper. They quickly crawl in at the hive entrance and the performance is over. For a few minutes many bees will of course take wing and fly over the hive and the novice may at this time be in great fear of being stung but there is really little danger even though it is well to have the smoker conveniently handy.

Sometimes the bees resent their new quarters and insist upon swarming out—absconding, the beekeepers call it. This can be prevented by providing the new hive with one or two frames of brood from the old hive body. The instinct to care for the young asserts itself in such a case and the bees "stick."

Commercial honey is produced and marketed in two forms known as "extracted" and "comb" honey. When an apiary is managed with a view of producing extracted honey swarming can be more easily controlled because an abundance of room can be given at all times.

The "supers" for producing extracted honey are usually regular hive bodies equipped with frames in every way similar to brood frames. As these are filled they are removed, the cappings sliced off and the honey extracted by means of a centrifugal machine which whirls the honey from the cells. The frames are not injured and are returned to the hive body for the future use of the bees. One set of such combs will be used time after time and in this way the bees are saved the labor of constructing the waxen containers for their product. ¶ Naturally a colony of bees will produce much more extracted honey in a season than it would comb honey. In the past few years many beekeepers have almost abandoned the production of comb honey because of the comparative ease with which the extracted product can be obtained and also because of the high prices prevailing. This has resulted in almost unheard-of prices for good comb honey and as might be expected many producers are now turning their attention to this form.

Comb honey requires more expert attention than the production of extracted because of the greater liability of swarming and also because of the difficulty sometimes encountered in getting the bees to go to work promptly in the supers.

The supers in which comb honey is stored contain the little square boxes each with its sheet of foundation. In the center of the super should be placed a "bait comb"—that is, a section containing some partly drawn foundation. These bait combs should be saved over from the previous season and are simply sections which the bees of the year before did not get filled with honey. They seem to act as a suggestion to the bees to let them know what is expected of them. The supers should be placed on the hive *before* the honey flow is expected. To delay this may result in the hive being so crowded that the bees will get the swarming fever and, although we control the swarm, we interfere with the actual working hours of the bees during the harvest period. Hives should be so managed that at the beginning of the honey flow the bees have every available space in the brood chamber filled either with brood (sealed if possible), or with honey. Then, having no space to store the surplus, they readily go to work in the supers. By adding additional supers as needed they can always be given plenty of room and will often stay on the job steadily as long as the flow lasts.

If, during the honey flow, the colony insists upon swarming, the scheme outlined above must be worked to induce the bees to remain at work *in the same set of supers*. Also the parent hive must be replaced with one containing, not empty brood frames, but frames fitted with sheets of foundation only. Perhaps *strips* of foundation are better than full sheets. The reason for this is that if we provide storage space and brood rearing space at such a time, the bees will give all their attention to these matters and allow our supers to be neglected and in consequence we may lose the crop. By giving the strips of foundation we certainly prevent brood chamber work as much as we can. However, in such an arrangement, we must prevent the queen from entering the supers lest our fancy comb honey be converted into a series of bee nurseries, for the queen would probably just as soon lay her eggs in the section boxes as in the brood frames. She is not at all particular about that. To prevent her from enjoying any such liberty a "queen excluding honey board" is inserted between the brood chamber and the supers. This is a board arranged with wires so spaced as to allow the passage of a worker bee but excluding the larger bodied queen. They are built in various styles and are a part of the stock in trade of all dealers in bee supplies.

It may be asked, "What becomes of the parent colony after we set it aside when the swarm issues?'' This hive should be turned at right angles to the new hive so that any field bees that are out in search of supplies will enter the new hive and not the old one. After two or three days the parent hive is turned part way back so that its entrance is closer to that of the new one. In a day or two the parent is moved still more so that it stands alongside the hive occupied by the ''swarm.'' A few days later the parent colony is opened and most of the bees are shaken in front of the ''swarm.'' Then the parent hive is moved entirely away and allowed to build up its strength either for a fall flow or for the next season. So the bees really do accomplish what they set out to do, namely, establish a new colony, but they do it in such a way that their owner reaps a profit from their work.

CHAPTER VI

How would you like to be the family physician to a colony of bees?

That is the position occupied by Doctor G. F. White of the Department of Agriculture—only Doctor White is a sort of physician extraordinary to all the bees in the United States. His job is to study the various diseases of bees and determine their cause and to investigate methods of cure and prevention.

Doctoring a sick bee is not quite such a fearsome business as it might appear to the beginner, however, because most of the serious bee troubles are not with the adults but with the brood, the larvæ. I suppose we might call them "diseases of childhood," and the treatment does not

consider the welfare of the patient but
only that of the colony. Consequently a
sick bee does not get any tender care from
the doctor—its name is Dennis right from
the start and the only fate it can hope for
is to perish as quickly and quietly as possible. I suspect that if the society for the
prevention of cruelty to animals and the
anti-vivisection agitators knew how many
sick bees were put to death every year
they would raise a howl to be heard in
Mars.

The two most serious bee diseases are
known as American and European Foulbrood, respectively. The names really do
not indicate anything regarding the origin of the diseases for both of them are
known on both sides of the Atlantic. Long
usage has established the nomenclature,
however, and by these names they will
probably always be known.

Both are diseases of the larvæ, as might
be supposed from the names, and both are
of bacterial origin. American foulbrood

is caused by an organism known as bacillus larvæ, while the other form is caused by bacillus pluton. Both of these organisms are exceedingly resistant to ordinary methods of disinfection. The first named is much more resistant than the second as it may be boiled in honey for thirty minutes and still live and will resist a five per cent. solution of carbolic acid for months. Material containing either bacillus may be dried and kept for at least a year and still be capable of causing the disease.

Both diseases are spread from one colony to another by means of infected honey. The trouble is not air borne, but must be carried from one hive to another or from some point of infection other than a hive to a healthy colony. Such points of infection may be jars that contain a residue of unused honey obtained from diseased colonies; diseased combs taken from colonies affected with the trouble or hives in which the bees have died or have become so decimated by disease that robbers

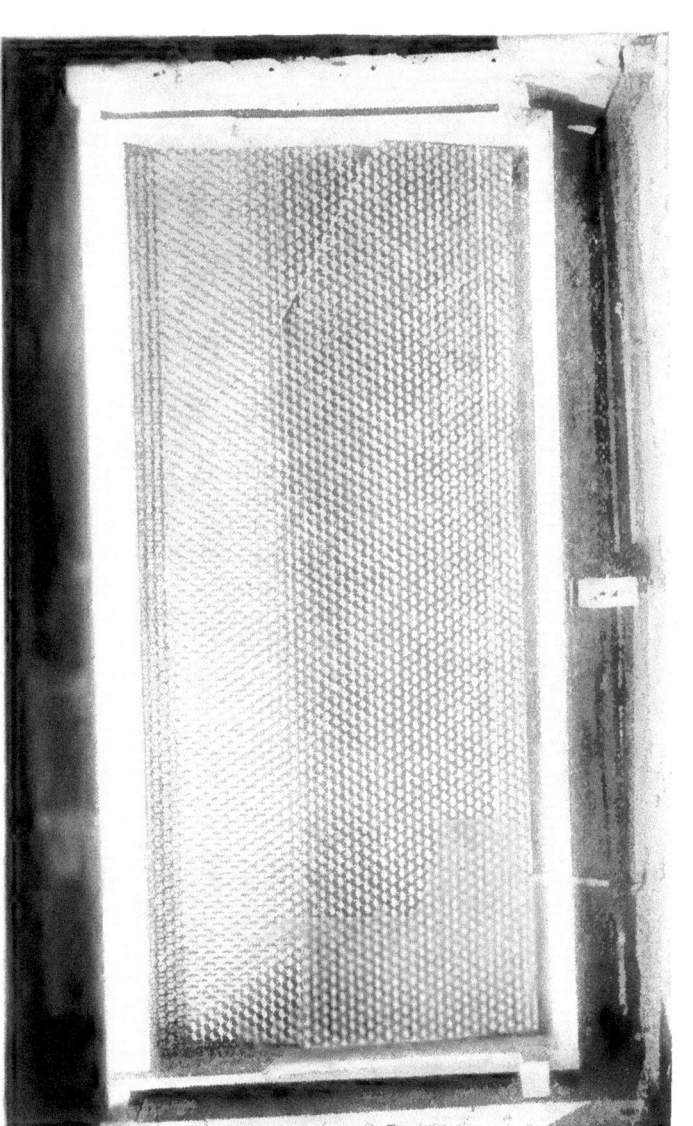

A brood frame containing a full sheet of "foundation."

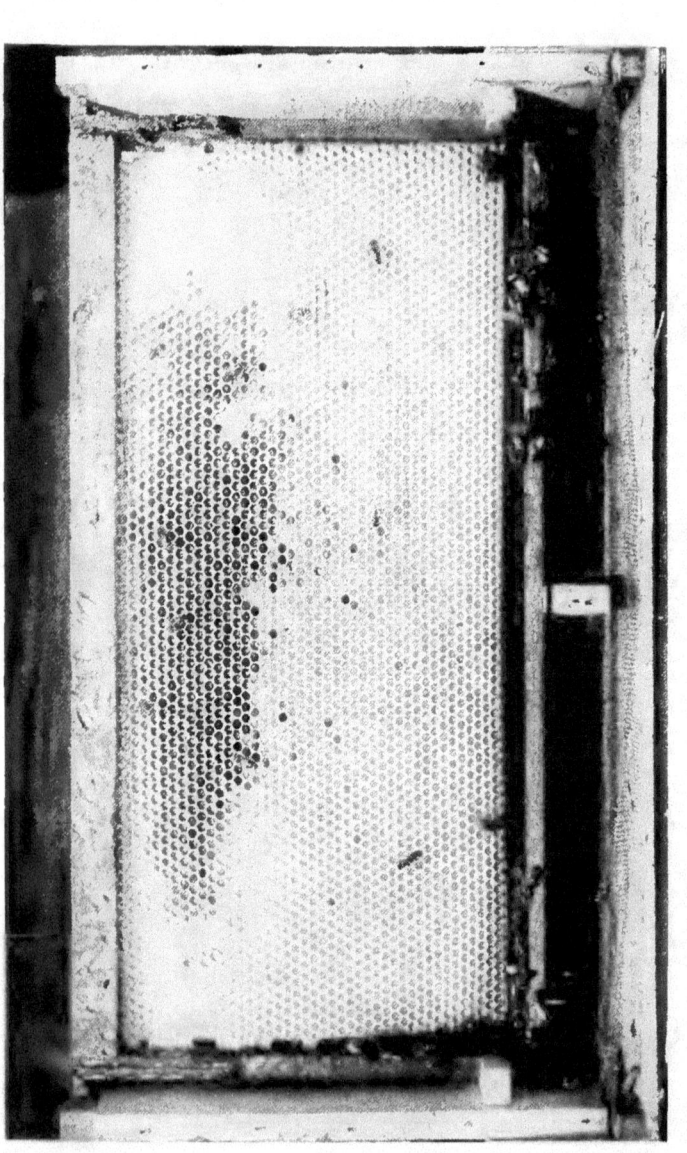

The bees have started to "draw out" the foundation, forming the cells in which honey is stored and young bees are reared.

can no longer be kept out. The tendency of the bee is to be thrifty and it usually carries this disposition so far that it does not hesitate to rob a colony that is too weak in numbers to defend itself. Consequently when a colony is attacked by either of these brood diseases there is a very great chance that healthy bees in the neighborhood may contract the disorder on account of their tendency to appropriate what does not belong to them.

Probably the first evidence that the bee-keeper will have to indicate disease in his colonies will be the dead brood. If American foulbrood is present a peculiar foul, sour smell will be noticed, but this is not evident in the case of European foulbrood except in certain very advanced cases.

The disease attacks the brood just before and during the period when the young bees are being sealed up to undergo their transformation to adult bees. The European form attacks somewhat earlier and

the American form is most in evidence a
trifle later than that stage. The larvæ in
each case becomes discolored, changing
from its clear normal white to various
shades of yellow and brown, eventually
becoming nearly black. In American
foulbrood the appearance of the cappings
over the sealed brood is quite characteristic as they appear at first sunken
and later will be perforated with small
holes. This condition is not observed with
European foulbrood, the cappings remaining normally convex or nearly level.
When a diseased larva is punctured with
a small stick such as a match, the contents
will be found to be viscid or ropy in the
case of the American disease. Later the
larva will dry down to a closely adhering
scale lying on the lower wall of the cell.
Such scales are removed with difficulty,
but in the European form the scales are
easily removed.

The treatment for either disease is
similar and consists in the removal of all

infected and infecting material from the colony. Many "systems" of treatment have been devised to secure this result but to mention all of them would only confuse the beginner and perhaps lead him to a lack of confidence in the use of any particular method.

The diseased colony should be "treated" during a period when the bees are gathering plenty of stores from the field. Probably the middle of the day is the best time to do the work as at this hour the field bees will be out gathering fresh supplies and the population of the hive will be accordingly less than during the late afternoon or early morning hours.

A fresh hive known to be free from disease is provided. The frames should contain only strips of foundation about an inch wide. Then we should have an empty hive body or other box provided with a lid which may be quickly and tightly closed. The diseased colony is gently set to one side and in its place is set the new hive. Spread newspaper in front of the hive so that the bees can crawl up to the entrance without having to claw their way through any grass or weeds in front of their new dwelling. Then as quietly as possible remove each frame from the diseased colony and give it a quick downward shake so as to deposit the bees on the newspaper. Repeat this as quickly as possible with each frame in the hive and as quickly as the frames are cleared of bees they must be placed in the empty hive body provided for that purpose and the lid should be replaced after each frame is inserted. As soon as all the bees have been shaken in front of their new quarters they should be urged to enter the hive by the liberal use of smoke. This too will tend to keep bees from other colonies from investigating the moving day program and hence will avoid any chance of some strange bee carrying a load of infected honey back to his healthy children. As soon as the bees have entered the hive the newspaper should be gathered up and burned.

In the case of European foulbrood, it is now customary to re-queen the colony at the time the swarm is shaken. It appears that the old queen is able to convey the disease to the new hive. There are various ways of introducing the new queen to the colony after the old one has been killed but this is work that the beginner had best leave to the hands of some experienced friend as the bees do not always welcome a new "ruler."

It will be noticed through this entire process that the object is to get the bees into new quarters and to avoid the chance of any outside bees carrying away any disease-producing honey. Of course the bees carry with them into the new hive all the honey they can hold, but since they are confronted with brood frames containing only strips of foundation they are unable to begin rearing any brood for some time. They must first draw out the comb in which the queen can lay eggs and in this interval they consume all of the diseased honey which they brought with them and the new brood, when it does appear, is fed with fresh honey recently gathered from the fields. The combs from the diseased colony which we so carefully protected by placing them in a tight box may be disposed of later. If only a few colonies are diseased the best thing to do is to burn the entire combs, frames and all. The hive body may be rendered perfectly safe for use by burning it out with a gasoline blow torch or by scorching the inside with burning gasoline so as to scorch the wood to a slight extent.

If many hives are involved in the treatment, the diseased brood frames may be placed in hive bodies and tiered up over some weak colony until the brood hatches. (Not all the brood is diseased of course, some of it is normal and will hatch.) In this way a weak colony may be built up to one of good size and it in turn may be treated as above outlined.

The essentials in the treatment are the removal of all infecting material and the protection of this material from robber bees. Sometimes the bees resent being transferred to a new hive and have a tendency to abscond. This may be prevented by placing over the entrance what is known as an entrance guard which is simply an arrangement of wires so spaced as to allow the passage of worker bees, but through which the larger bodied queen can not pass. If the bees swarm out past such a guard, they quickly find that the queen has been left behind and they then return home and are usually satisfied to go to work.

If a diseased colony is neglected it will die sooner or later and any honey remaining in the hive will be carried out by the other bees in the neighborhood. Sometimes bees will rob colonies several miles from their home apiary and in this way disease may be spread.

Consequently it is to the interest of all beekeepers to cooperate with one another in an effort to stamp out these troubles. One man working alone can not do much to keep his bees healthy but all beekeepers in a neighborhood working in cooperation can rid their community of disease and keep it in that condition. Many states have bee inspection laws and proper officers to enforce them but laws are of little effect unless the public is willing to cooperate with the law enforcement officials. As long as the public smiles on "boot-legging," prohibition will not prohibit, and as long as a careless beekeeper insists upon keeping bees in inspection proof-box hives and as long as he fails to treat diseased colonies our bee laws will be of little avail in preventing the spread of disease.

Another brood disease that is not so well understood as the two foulbrood diseases, is called "sacbrood." In the past many different names have been applied to this trouble which is characterized by the death of the larva and the formation of a scale on the side of the cell. These scales (the dried bodies of the larvæ) always turn up at the tip, giving what bee-keepers have long termed the "Chinaman's shoe" effect.

Like the other brood diseases it is infectious though probably less so than the first two disorders mentioned. Doctor White has done much admirable work on this disease but has been unable to isolate any definite organism which can be labeled as the actual cause of the troubles. He has found, however, that the disease can be transmitted by means of the dead larval remains. Such remains contain a virus of which very little is known except that it will reproduce the disease. It is apparently not found in honey. Consequently the danger of transmitting the disease from one colony to another is much less than it is in the case of either American or European foulbrood.

Colonies affected with sacbrood also show a disposition to recover from the disease so that it is not quite such a hopeless disorder as are the first two troubles.

Diseases of adult bees are not so well understood as are those of the brood. Bee paralysis is a condition that has been observed by many beekeepers, but in most cases it is not a serious or wide-spread trouble. It is probable, too, that the term has been applied to more than one disorder.

Nosema-disease is probably wide-spread. It is an infection of the intestinal tract of adult bees and may be transmitted from one colony to another. The affected bees exhibit no particular symptoms but seem to be able to attend to their functions about as usual. Colonies affected with the disease are nearly always weak and show less resistance to unfavorable winter conditions than do healthy colonies. Probably much of the trouble described by beekeepers as "spring dwindling" may be traced to this disease. However, it is a trouble that requires expert knowledge to diagnose. Some work has been done along the line of developing a medical treatment by the use of drugs in this disease, but the results so far are not conclusive enough to warrant any definite statements as to what may be expected along this line.

Dysentery is a disease of the late winter months caused by the fact that the bees have been closely confined to the hive and retain in the intestinal tract the accumulated fecal matter of several months. During prolonged cold weather this may result in the death of many individuals of the colony. Ordinarily the trouble corrects itself if the bees have a few hours' flying during some warm day. At such times the hive and the surroundings may be liberally marked with brown. During open winters when the bees are able to fly every few weeks the trouble seldom appears. Doctor L. Bahr, of Copenhagen,

Denmark, has recently called attention to a condition which he calls "bee typhoid." He has described a bacillus which he thinks is the cause of the trouble. Doctor White in commenting on the Danish discovery says that we probably have the trouble in this country and it more than likely is being included under the general term "dysentery."

It should be said that none of these bee diseases could possibly be transmitted to man. Those which are produced by known bacteria may have the causative organism present in the honey which we eat, but we can as entirely disregard them as though they were non-existent.

Bees have other enemies besides the diseases mentioned and frequently these enemies go hand in hand with the more serious diseases. Of these the wax moths are by far the most spectacular. I have on many occasions heard beekeepers say that the "wax moths" had killed all or part of their colonies. Such a statement

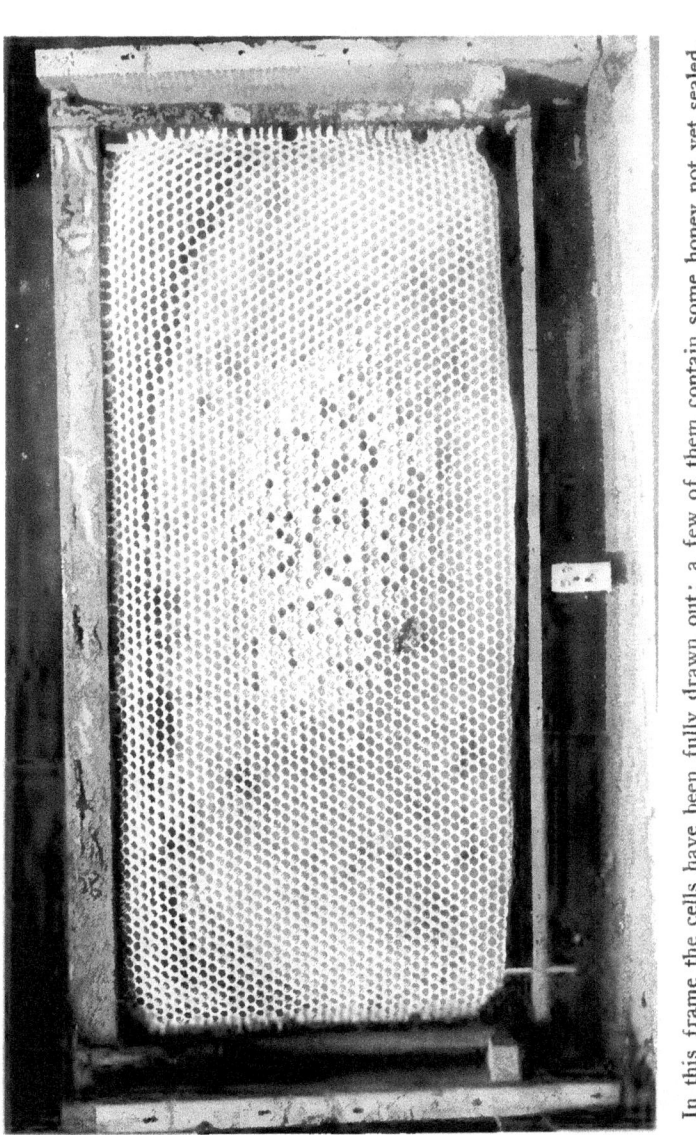

In this frame the cells have been fully drawn out; a few of them contain some honey not yet sealed. In the center of the frame is a patch of "brood" with a few cells capped. The single bee shown is the queen.

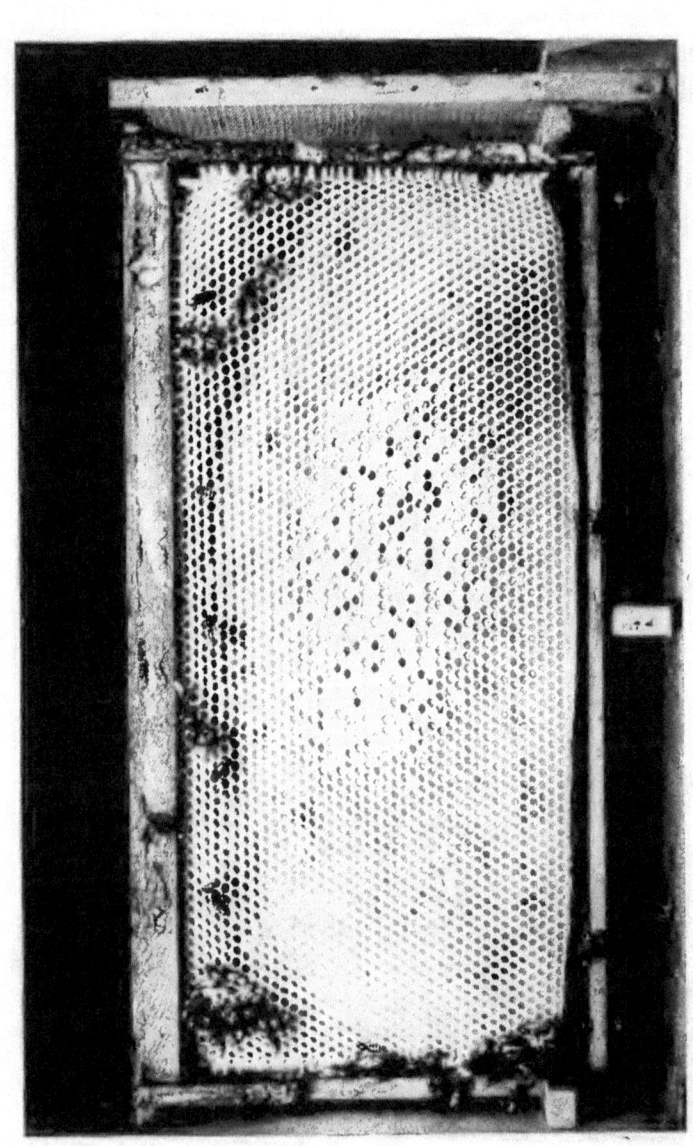

This is the same frame shown in the preceding illustration as it appeared a few days later. There is more honey and the area occupied by brood has increased.

always shows ignorance on the part of the beekeeper because the work of the wax moth is almost invariably of a secondary nature. In a strong colony a wax moth would have a hard time doing much damage because the watchful inhabitants would hustle it out at the front door almost before it emerged from the egg.

If the colony is weakened from disease or neglect the wax moth soon ruins the combs and puts an end to the colony. The eggs are laid in the cracks and in the entrance to the hive. As the young hatch, they crawl to the combs where they feed and spin their webs. In the end the combs are reduced to a tangled mass of débris consisting of the hard portions of the comb combined with the almost spider-like webs which the feeding "worms" produce. They work with almost unbelievable rapidity when they once get a start. I recently saw a hive containing empty combs which had been left in the apiary for about two months and all but one or two of the brood combs had been entirely destroyed and the remaining ones ruined for any future use.

Whenever they are found we may be quite sure that some other agency is at work to weaken the colony; the moth being only a danger signal to warn us to look closely and discover what it is that is weakening the bees to the point where they are no longer able to resist invasion.

In some sections of the country ants are a serious pest in the beehive. They may be prevented from reaching the colony by using a low platform set up on short legs. Each leg should be stood in a tin cup containing any heavy oil. The ants will not cross the oil. Such an arrangement must be given occasional attention, however, for rain-water will collect in the cups and float off the oil. In place of the oil the legs of the hive stand may be banded with tree tangle foot, a sticky preparation similar to that used on fly-paper. It will remain viscid and effective during the length of the average summer and will effectively prevent ants and other insects from crawling into the hive.

With the exception of American and European foulbrood all of the pests of the apiary can best be combatted by keeping all colonies in a strong condition. The natural resistance of a strong colony is of more service to us in repelling obscure and little known diseases and insect enemies than any amount of fussing about the hive that we can do. The two forms of foulbrood are just as liable to attack a strong colony as a weak one, and in such cases our assistance is imperative if we are to save the life of the colony. For most other troubles our best course is to manage our bees so that at all times the colonies are strong in numbers and are provided with sufficient stores. Under such conditions the honey-bee, especially if it be of the Italian race, can nearly always hold its own.

CHAPTER VII

HONEY PLANTS

BEGINNING beekeepers, and occasionally "old timers," ask what flowers they can plant in order to provide pasturage for their bees. Too often they have the idea that a few extra pots of geraniums or a couple of hollyhocks will make a difference in the honey crop. They fail to realize that plants suitable for bee pasturage must be exceedingly numerous before any perceptible results may be noticed in the amount of honey that is stored in the hive.

Some beekeepers have made a practice of planting an acre or two of such plants as buckwheat, thinking that they will thereby furnish a regular fall feast to the honey gatherers. It is true that buck-

wheat does yield nectar in tremendous
amounts, but an acre or two would be only
a drop in the bucket for the average apiary. One must realize that a colony of
bees contains many thousands of individuals, each of whom ranges over an extensive territory and carries home only
an infinitesimal amount of honey making material.

Consequently any honey plant must be
one that not only produces quantities of
nectar but which also occurs in sufficient
numbers to provide the necessary bulk of
nectar. Comparatively few flowers do
fulfil those two conditions although a
great many serve to supply the bees with
moderate amounts of nectar and thus help
to tide them over the lean portions of the
summer.

Throughout the East white clover has
long been the leading honey plant. At
one time the yield from white clover probably exceeded that from any other plant,
but with changing conditions it has be-

come less common. Formerly every pasture contained patches of this clover so dense as to constitute almost a "pure stand," but with the better understanding of crop rotation the old pastures have been broken up and seeded to other crops. As a result the white clover has been driven back to road sides, fence corners and abandoned fields where it still thrives and still produces limited amounts of its beautiful honey. The plant is not a native but was probably introduced from Europe or Western Asia very early in the settlement of this country. It is claimed to be identical with the "Shamrock" of Ireland although this is another of these matters which are in dispute among Irishmen and far be it from me to express my opinion.

The yield from white clover is often dependent upon the weather conditions to a very great extent. Being a biennial it must grow from the seed one season and bloom the next. Consequently, cool moist

summers are favorable to the growth of
new plants for the next year; but if we
have cool moist weather during the blooming period the secretion of nectar is re-
tarded and the consequent honey crop is
short. On the other hand, warm weather,
and particularly warm nights, during the
blooming period tends to promote nectar
secretion and the crop may be expected to
be large in proportion. Such secretion,
however, may be cut off entirely if the
weather is too hot and dry, for under
these conditions the period of bloom is
very short, the plants dry up and die and
the white clover honey flow is over for
another year. All of these factors have a
very important bearing on the management of the apiary and show how necessary it is that the beekeeper have his colonies ready to gather the white clover
nectar promptly. Sometimes the flow
will last for weeks and again it may be a
matter of a few days and the beekeeper
who is not ready for it when it comes will
fail to reap any profit.

The same thing is true, in a lesser degree perhaps, with most other honey producing plants. It is important that the beekeeper know from what plants his surplus is most liable to come and then be prepared to secure that surplus promptly when the time arrives.

The introduction of sweet clover, an European weed, into the white clover sections has had something to do with the crops from the older plant. Both bloom at about the same time and as a result the fine white clover honey is contaminated with the greenish colored product of the sweet clover. While the latter is by no means a bad honey it does not rank with the white clover and the mixture has been the means of practically eliminating the older honey from the market.

Perhaps I am a "crank" on white clover honey. Certainly I am very sensitive to the least trace of strong honey when it is mixed with white clover. Some curious arrangement (probably a defect) of the nerves of my throat enables me to detect instantly any admixture of foreign honies when combined with the pure clover. Clover and basswood honey I can eat in unlimited amounts, but most other varieties cause a sensation similar to what I imagine I would get if I tried to chew a mouthful of nettles. Sweet clover, when mixed with white clover honey, causes this unpleasant sensation in a minor degree and this is one of the reasons why I have never been friendly to the introduction of this new plant. Each year it is more and more difficult for me to secure honey for my own use—I have long since given up trying to produce it in my own locality. When pure white clover honey was a standard product one could count on a crop of palatable comb every year, but—"them was the childhood days."

Alsike clover has been widely planted as a forage crop to take the place of the red clovers usually grown for that purpose. In the sections where it will grow it should be more widely planted as it produces a better quality of hay and often a larger quantity than does the red clover. It is grown in the same way as red clover and in soils adapted to its culture will live from year to year and about as long as the red. On dry soils it does not succeed and soon "runs out" leaving the field to weeds. Many beekeepers have adopted the plan of furnishing alsike seed to their neighbors at less than the market price in order to get them to plant large acreages of it, and where this has been done a noticeable increase in the honey crop has been realized. The honey produced is very similar to that from white clover, and probably many consumers will fail to distinguish one from the other. It has become an important honey plant in many sections and should become still more important each year.

Ordinary red clover yields large amounts of nectar. Perhaps I should say

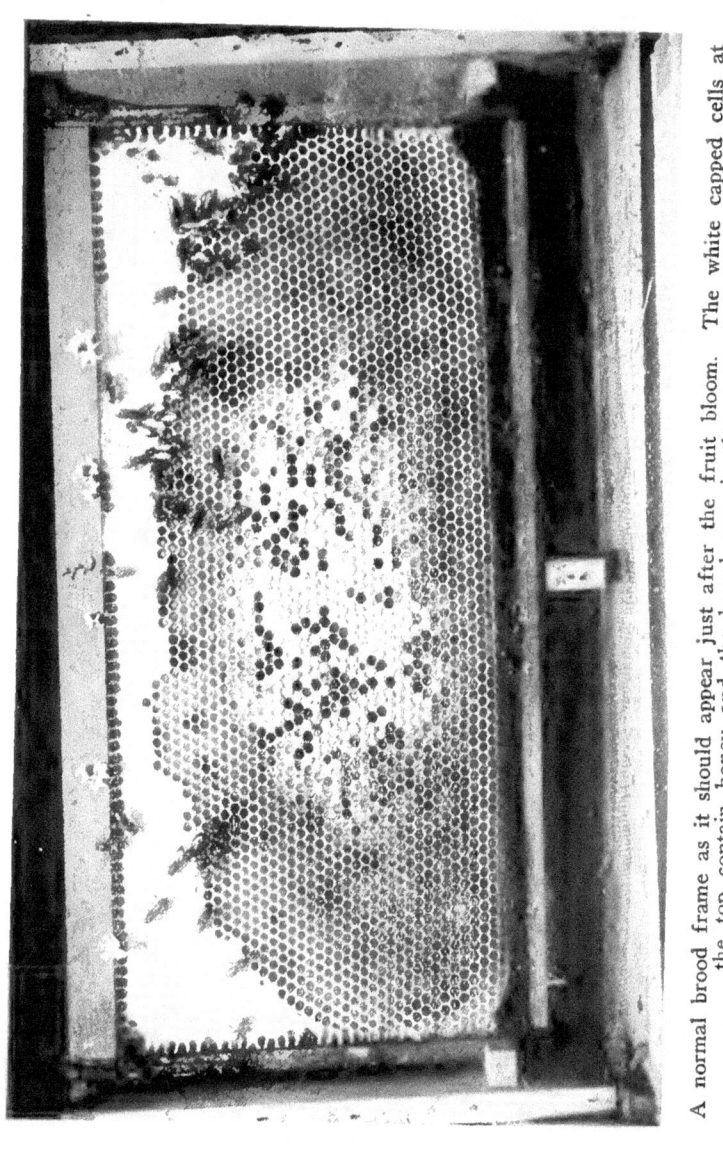

A normal brood frame as it should appear just after the fruit bloom. The white capped cells at the top contain honey and the brood area in the center is increasing.

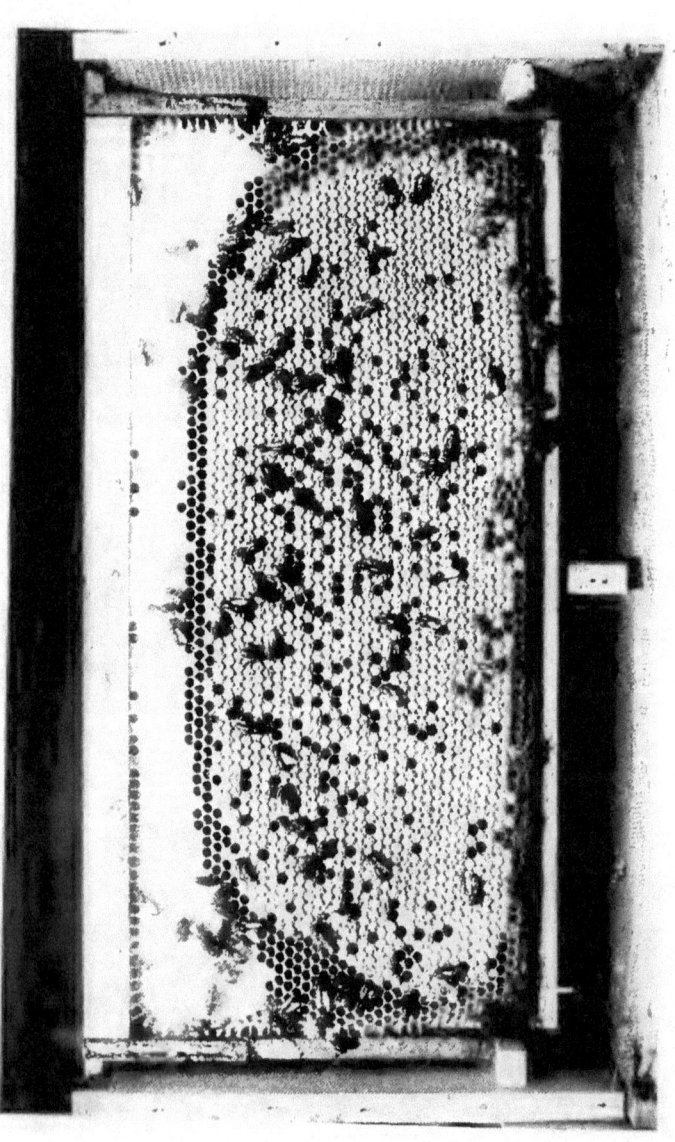

A typical brood frame as it should look at the beginning of the main honey flow. All available space is occupied by brood or stores.

it secretes large amounts of nectar because the flower tubes are so deep that the
bees can not reach the nectar content.
Consequently we lose tremendous amounts
of honey every season simply because the
bees are not built to red clover measure.
For a long time it was the dream of certain bee-breeders to develop a race of long-
tongued bees that would be able to harvest this crop of honey from red clover.
Such a development would be of very great
advantage to the industry but so far it has
never been brought about. Some seasons,
due to hot dry weather, the flowers of the
red clover are so stunted in size that the
bees appear to be able to reach the nectar
and this fact has no doubt helped to support the claim made in some quarters that
a long-tongued bee had actually been
produced.

Crimson clover, another forage plant of
value, has been considered an important
honey-yielding plant. It is planted late
in summer and produces flowers early the

next season. The bloom comes between the fruit bloom and that of white clover and for this reason is rather important in white clover sections. Ordinarily there is a long gap between fruit bloom and clover which does not furnish the bees with as much forage as they should have in order to enable them to build up their colonies. With more general planting of crimson clover in the white clover districts this gap would be effectively bridged and while a great increase in the crop might not be expected we would at least find our bees in better position to give an account of themselves during the heavy white clover flow.

In parts of the West, alfalfa, another important clover, has become a great factor in honey production. The honey from it is not quite so fine as the best white clover honey, but it ranks near the top in the matter of quality and always finds a ready market.

It should be remembered that alfalfa is

usually cut for hay just as it comes in bloom and as a result the large hay ranches do not furnish the bee pasturage that might be expected. It is only where the plant is allowed to ripen its seed that great returns are obtained and in these sections apiary locations are often at a premium—so eagerly are they sought.

There is no reason why alfalfa should not occupy a more important place in eastern agriculture than it does to-day. In parts of the West and Central West it is of great importance and to a lesser extent it should become important in the East. Beekeepers can do much toward this end by urging the planting of experimental plots among their friends in sections where the plant is not already being grown.

Among our native plants none ranks higher than the American linden or basswood. It is recorded as having furnished nectar for a greater per day production than any other plant. One colony has been known to store over twenty pounds of honey in one day from basswood bloom. Mr. G. M. Doolittle, a pioneer in scientific beekeeping in the United States and a man who has contributed much to our knowledge of the honey-bee, states that he has never known the basswood to fail to yield a good supply of honey. The flow, according to him, lasts from three to nearly thirty days. Unlike white clover the yield from basswood is not liable to be interrupted by rains. The flowers of this tree are pendant and hard dashing rains do not remove the secreted nectar at the base of the petals. Like all other flowers, however, it yields most nectar during periods of warm nights.

Mr. Doolittle's observations cover a long period of years—a period during which the basswood was a far more common forest tree than it is at present. Originally it was one of the fine timber trees of our American woods, but because of the value of its timber, it was eagerly

sought by the lumber men and it has
now vanished from the greater part of
the country it once occupied. Sections
which one yielded a reliable crop of
basswood honey now do not produce a
drop of this delectable product simply
because the tree has been all but ex-
terminated. The wood is soft, light and
easily worked and all of these qualities
caused it to be very much in demand. One
curious thing about the vanishing of this
fine tree is that the beekeepers themselves
have played no small part in its destruc-
tion. For many years the little wooden
sections in which comb honey is produced
have been made from basswood exclusively. In fact I do not know that I ever
saw a section made from any other wood.
This use may be thought to play a tiny
part in the consumption of this kind of
timber, but in the aggregate immense
amounts of basswood have been thus used.
To-day, I know of no place where this
tree is sufficiently common to make it an

important honey plant unless it be in parts of Michigan and Wisconsin. Several years ago in Vilas County, Wisconsin, I saw great forests of hard wood trees in which there were a great many lindens. At this time the hungry hordes of timber companies were making inroads at a rapid rate and this last outpost may be gone to-day. The tree is of sufficient import in honey production that the beekeeper can well afford to take the time to plant it along his roads and around his home. It makes an admirable shade tree, blooms at an early age, and as it grows older becomes an important pasturage factor in the life of the bee. Ten trees, twenty years old, ought to supply a colony of bees with enough nectar to enable them to store a considerable surplus of honey and while some of us may not be here twenty years hence it is pretty certain that some one else may and it is possible too that "some one else" may want to keep bees.

Another native tree that once yielded

large amounts of honey was the tulip or yellow poplar. The great cup-shaped flowers sometimes contain an abundance of nectar from which the bees make a strong dark honey, unsuited for human food but of value in brood rearing.

Like the linden, the tulip-tree is rapidly reaching the point where it is almost a curiosity. It was one of our finest forest trees and will be missed more by the tree lover than it will by the beekeeper.

In some parts of the country the smooth or black sumach yields tremendous quantities of honey. It is a plant of dry hillsides and is often found in those portions of the low hill country where clover is scarce and where the native basswood has mostly been cut out. In such situations it sometimes gives the beekeeper a surplus in years when all other plants fail to make a showing.

Any one who has ever visited the cut-over portions of northern Michigan or Wisconsin in the summer is familiar with the "fire weed," that pink flowered wanderer that takes possession as quickly as the fire has made a place for it. The tall stalks, sometimes reaching a height of six feet, are topped with a cluster of bloom that grows with the season and unfolds a succession of flowers from July until frost. In many of the burned areas it is the dominant plant for years after the fire has killed the young trees that were left by the woodmen. Thousands of acres of such country exist along our northern boundary and south of that line, and in the whole territory there are comparatively few bees kept.

And yet, the honey that the bees obtain from the fire weed is nearly as fine as that from white clover. Aside from possible trouble in severe winters it would appear that this burned area offered about the last stand in the way of virgin pastures for bees in this country. I am personally familiar with some thousands of acres of such pasture in which I know to a certainty that not a bee is kept and as the plant yields large amounts of nectar it appears as though some one were over-looking an opportunity.

The wild raspberry of the northern states contributes a very large share to the national honey crop. The drooping bloom which opens in early summer (late May or early June) resembles the linden in that the nectar is not easily washed out by rain. This drooping character of the bloom of these two important honey plants may account for the fact that both of them are reliable yielders practically every year. The raspberry honey is white in color and has a delightful flavor that easily places it commercially with white clover—the inevitable standard of excellence among all men who produce honey.

In the South much honey is collected from the bloom of one of the hollies. This low growing, evergreen bush ranges on low flat lands from Massachusetts to

Florida but reaches its best development in the southern part of its range where it is called "gall berry." It sounds like a strange name for a honey-producing plant, but nevertheless the product is said to be entirely free from any tinge of bitter. It is a honey with which I am not personally familiar but I am told that it is of fine quality and that it is produced in tremendous quantities. Unfortunately beekeeping in many parts of the South is in a very backward state and comparatively few really modern apiaries exist—certainly not nearly so many as the flora of the country could seem to justify.

In many parts of the country the "fall flow" is the important source of the commercial honey crop. In these sections the beekeeper must manage his colonies in such a way as to have them at their maximum strength just at the time when this flow is to be harvested. In such a district it would be the height of folly to build up strong colonies in the spring and perhaps be forced to feed them over the hot weather—during which there is nearly always a scarcity of bloom in all sections.

The varying swamp flora furnishes a tremendous yield of nectar and is eagerly gathered by the bees.

Buckwheat, cultivated in some northern states for its grain and in some places as a cover crop, produces a dark honey with a very characteristic flavor. It is relished by many who have acquired a taste for it, but to one who was brought up in the Middle West and who has been familiar with white clover honey from the days of his infancy, the product of the buckwheat offers small appeal. I remember an old New York farmer who moved "West" many years ago always complained because he could not buy any real honey. "This stuff you have in your stores out here," he said, "is all manufactured honey. Why can't your grocers get me some real honey?" I offered to supply him with some "real honey" and presented him with a section of my best white clover product. It required a great deal of argument on my part to convince him that I also was not in "cahoots" with the food adulterators. As a matter of fact there has never been a pound of artificial comb honey produced by any one at any time. A high government food expert once made the break of mentioning artificial comb honey in one of his reports and he has been the laughing-stock of beekeepers from that day to this. Mr. A. I. Root, who might be called the dean of modern beekeepers, offered a thousand dollars reward for a single pound of artificial comb honey. This reward was offered many years ago; it still stands but has had no claimant.

But this is a digression, justified only by the fact that to many folk buckwheat honey is a standard for comparison just as white clover is for others. The two honeys do not resemble each other in the

A brood frame late in the season after brood rearing is over. There is an abundance of food stored in the white-capped area at the top of the frame.

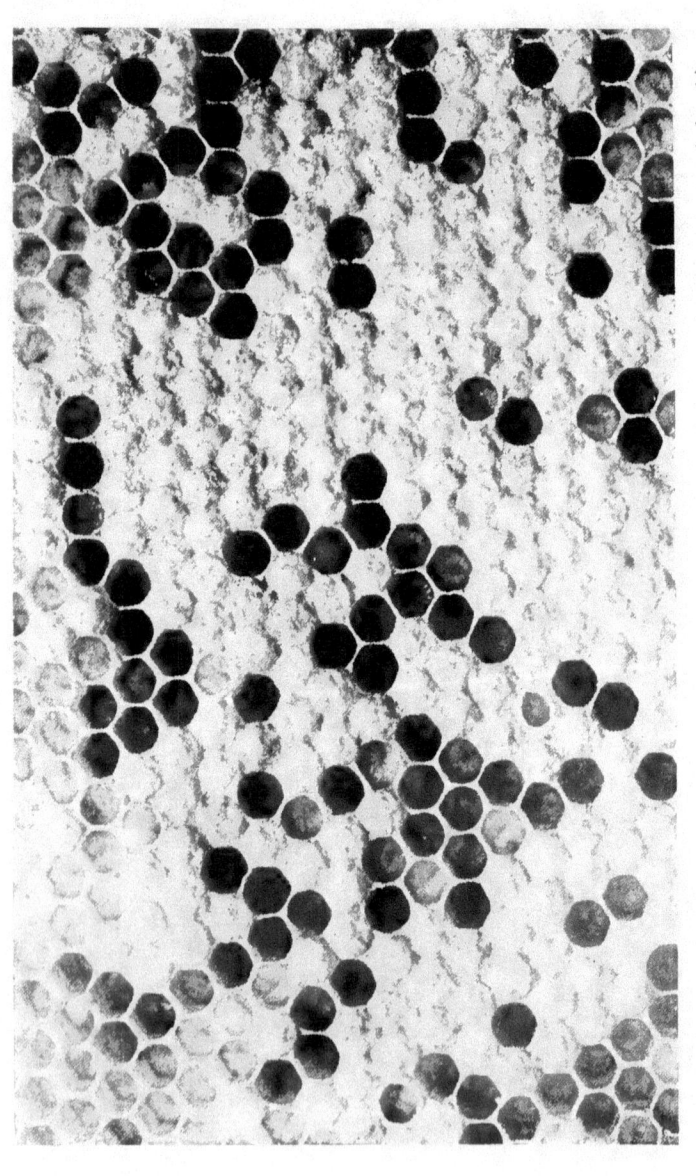

A section of brood comb photographed natural size. At the edges may be seen the curled white larvæ. Most of the cells are capped; they contain young bees nearly ready to emerge.

least and I can readily understand that a person accustomed to buckwheat would find the light and delicately flavored clover rather insipid.

While the buckwheat yields tremendous amounts of nectar I would hardly advise any one to plant it with the sole aim of securing the honey crop. In sections where it can be grown for grain the bee-keepers reap the incidental harvest of the honey. This process might easily be reversed by planting the grain for the benefit of the bees and considering the grain as incidental to the honey produced. In any event I would not give the impression that small acreages would have any influence on the resulting crop. The grain is not grown in my own part of the country, but one year we planted about five acres in a part of the orchard to act as a winter cover. It bloomed heavily and the bees worked on it but I never saw any surplus in the supers.

Related to the buckwheat is the heartsease, a member of the smartweed family. It grows in swamps and in wet places near stream banks. Often it occurs to such an extent that profitable yields are obtained and in a few locations tremendous surplus has been recorded. The honey is light amber in color, stronger than clover but still mild enough to find a ready market. A great variation in the honey is liable to occur because there are several varieties of the plant and other honey is liable to be mixed with it.

Asters of many kinds, goldenrod and numerous sorts of "wild sunflowers" furnish the bulk of the fall honey. Many of these plants live in low wet places and the northern Indiana marshes have long been a favorite location for commercial beekeepers. With such a wealth of flora to draw upon it is to be expected that there will be a blend from many flowers. For this reason we seldom ever find samples of "pure" aster, goldenrod or sunflower honey. All of them furnish a dark colored product, strong in flavor and not to be compared with the beautiful white honey from other plants. However, this dark honey finds a market and often the quotations are but little below those for clover.

Speaking of dark honey reminds me that a word should be said about "honey dew."

Frequently we find in the hives a dark strong substance that appears to be honey but that obviously is not. The bees gather it and store it exactly as they would the product from the flowers but they get it from the plant lice. Honey dew is nothing more or less than an excretion from the "honey tubes" of the aphids that feed on many plants. Some seasons the bees will gather tremendous quantities of this stuff and load up their storage space pretty liberally. It serves them well for brood rearings in the spring, but when it is gathered late in summer and used as winter food it often causes serious damage.

Honey dew can be eaten—at least I have known persons who ate it and said they liked it. As for myself—with that honey-sensitive throat I mentioned—if I should take one taste of honey dew nothing would help me but a fire extinguisher. The pure food laws specifically mentioned the product as not being saleable under the name of honey.

There is another class of honey plants about which the beekeeper should know and these are the very early blooming sorts that appear in spring. Practically none of them produces surplus, but many of them are valuable because of the fact that they enable the colonies to increase in strength in time for the later flowers.

The earliest blooming plants furnish practically no nectar but they do furnish what is quite as important to the bee, and that is pollen. Some of them, like the hazel, furnish pollen in tremendous quantities. Following the hazel bloom, the willows and true poplars supply pollen in al-

most unlimited quantities. The dande-
lion, that lowly source of "spring greens,"
cursed by the lovers of trim lawns, furnishes pollen in abundance and often at a
time when the bees are needing this ad-
junct to their nursery menu. Without
pollen brood rearing in the hive would
have to cease so it is quite as important
that the bees have access to it as that they
later have access to a source of nectar.

The bloom on the fruit trees is by far
the most important source of the early
spring honey and some years it is so lavish that the bees store more than they will
need. We have often been able to remove
a small surplus of pure apple honey and
found it to be of good quality though a
trifle darker than the best clover. Usually it is mixed with the honey from
plums, peaches and other tree fruits and
the resulting blend, to my notion at least,
is not so good as pure apple. All of the
honey from the fruit bloom, however, has
a delightful aroma—better in the nose

than in the mouth. It reminds one of those warm rich days when the plum orchards bloom and we get a whiff of sweetness across the hills on the evening air.

So important is the fruit bloom to the beekeeper that in sections where he has been depending upon it he must make other provisions in years of failure. If the trees fail to bloom or if the bloom is killed by cold, it is almost a certainty that the bees will have to be fed before the early summer flowers arrive.

Altogether there is a rather close partnership between the bee and the flower. Each is dependent upon the other to a very great extent—that sort of dependence which is extremely common in many of nature's works. That the beekeeper should understand this dependence goes without saying for no one can successfully manage bees unless he also knows something of the flowers which furnish the basis for his work.

CHAPTER VIII

BEEKEEPERS annually sustain a heavier loss from cold than is generally supposed. In fact the loss is probably much greater than even the beekeepers themselves realize. The actual colony loss for the United States is said to be more than ten per cent. and it is thought to be no exaggeration to say that the actual bee loss is near fifty per cent.

The fact that a colony of bees is alive in the spring really means very little as far as success is concerned. There may be enough bees in the hive to conduct business as usual but they will not be numerous enough to prove profitable to their owner when the honey flow comes on.

Really successful wintering means, or

149

should mean, that the bees live through with but little loss of individuals and in a strong healthy condition so that they will start promptly to rear the young bees for the next honey harvest. It is understood, of course, that all of the old bees that winter over die before the next active field season arrives. They serve only to start the new generation, but unless they are strong in numbers as well as in body they will fail in this very important function.

The subject of proper wintering has perhaps produced as much discussion as any other one phase of beekeeping—and that is stating the case about as strongly as possible, for beekeepers are prone to discuss their troubles and success with all the avidity of an invalid at a health resort.

Through all of this historic discussion run a great many references to the "winter sleep" of the bees. They are spoken of as animals which become dormant in the cold season and pass into a hibernating state similar to that enjoyed by the bears and other animals which curl up under an old stump and lose both consciousness and sense of activity until the winds of spring stir them into wakefulness. As a matter of fact bees do not become dormant in the winter. Quite the contrary. At all times they are active in the hive and strange as it may seem, the colder the weather the more active they become.

In all things heat is required to maintain life. In the case of bears, which have always been used as the type when we think of hibernation, this heat is provided by the excess stores of fat which are accumulated during the summer and fall. As the winter wears on the fat wears off, being consumed by the animal in generating heat. If bees hibernated they would have to be provided with some means of storing a reserve food supply in their bodies which would serve the purpose of the bear's fat—and a fat bee is still unknown to the fraternity.

Consequently the bee's heat generation is dependent upon what the bee has for breakfast throughout the winter. His food is more than nourishment, it is fuel—just as much as coal is fuel for a furnace. Like coal, too, it may vary in quality and burn with a clean flame and few ashes, or it may smolder and form clinkers, in which case the furnace or the bee will suffer.

With the consumption of this food,—or fuel—the bee combines great bodily activity in the form of exercise to keep warm, in very much the same way and for exactly the same reason that a man walks rapidly and slaps his body with his arms on a bitter cold day.

There are several factors which go to make successful wintering but they may be reduced to three primary ones, shelter, food, and the strength (numbers) of the colony. It is these three factors that we will consider in discussing the problem of making the bees live over winter.

Two brood frames filled with honey. The caps will be shaved off and the honey extracted by centrifugal force. This is the method used in obtaining what the public knows as "strained honey."

The larvæ of the bee moth have reduced this comb to a mass of débris.

Probably this question of shelter has caused more thought and more work on the part of beekeepers than any other one item. For a time after the introduction of the Langstroth hive with its movable frames it was a favorite indoor sport for beekeepers to design new types of bee homes. The idea back of most of them, however, was convenience in handling the bees rather than one of better shelter during the winter. Later attempts were made to build hives that would offer more perfect protection for the bees in winter but the first of these met with an indifferent reception. Probably even the best of the insulated hives offer but little advantage over those with single walls.

In a wild state bees live in the trunks of hollow trees with what appears to be only indifferent protection. Such colonies, however, seem to winter over in good condition. At least the colony seldom dies as a whole regardless of whether the loss in numbers is large or not. Probably, too, if we had any means of knowing exactly the physical condition of such a colony we would find that is was not quite up to our ideals so far as commercial beekeeping is concerned. The beekeeper desires more than to have his bees live over winter. He wants them to live over and to be in such perfect condition that they will store for him a surplus of honey —otherwise he might as well keep a colony of ants.

It is probable too that the protection afforded by a tree trunk is really greater than it seems. The wood walls are often several inches thick and the corky layer of bark furnishes an insulation of great efficiency. The same is true of the old straw "skeps" which were formerly used to house bees and which are still very largely used in some European countries. These skeps being made of twisted strands of rye straw contained in their walls innumerable dead air spaces which prevented a rapid loss of heat from the inside.

The modern single walled hive offers but poor protection to the colony in very cold weather. The loss of heat must be rapid and as a result the bees must be more active and must consume more food in order to replace the heat that goes out through the hive walls. This winter activity of the hive is an interesting thing and should have a more detailed description.

In bee literature many references will be found to the "winter cluster" and the impression has been that this winter cluster is a more or less immobile ball of bees formed for the protection of the individuals against cold. As a matter of fact it is far from being immobile, but instead represents the greatest activity. The cluster is not formed unless the temperature of the hive falls below fifty-seven degrees. At higher temperatures than this the bees remain distributed over the combs in a very inactive state—probably more nearly "dormant" than at any other time in the winter. When the temperature falls to fifty-seven degrees, however, they begin to bunch up and form the "winter cluster."

The outside bees on the cluster act simply as an insulating layer which helps to retain the heat generated by the bees in the center of the cluster. As the outer layer becomes chilled and stiff with cold, other bees crawl out from inside the ball and take their places—they in turn being relieved when they become cold.

The interior of this mass of bees represents ceaseless activity, ceaseless muscular exertion, and this exertion must have food on which to subsist. Consequently the bees inside the cluster must eat and they eat in proportion to the amount of work they are doing. The lower the temperature, the greater the activity and the more rapid the consumption of stores. Also the greater the cold, the more closely do insulating bees crowd together to protect heat generators inside the cluster.

From this it will be seen that the colder the weather and the poorer the hive the greater, will be the consumption of food. If we could keep the hive temperature above fifty-seven degrees no cluster would be formed, there would be but little activity and as a result only a slight amount of food would be used. From all of which discussion it is easy to see the inter-relationship between the hive, the food, and the size of the cluster.

With the knowledge that the single walled hives offered little protection to the colony there have been many attempts to devise double-walled hives of various sorts. Some of these have their advantages but in most cases the advantage is theoretical rather than practical. In sections where zero weather is common in winter it is doubtful if the double walled hive is of very great value unless it is given additional protection.

Beekeepers do not always agree as to how much or what sort of additional protection is necessary but many experiments have indicated that it is a practice not likely to be overdone.

In experiments by Demuth and Phillips, of the Department of Agriculture, it was found that one of the most convenient ways of packing was to place four hives together in a large packing case, leaving little tunnels to the entrances. This packing case was insulated by a layer of three inches of material below the hives, five inches on the ends, six inches on the sides and from eight to twelve inches on top. They state that the amount of packing used was effective for the latitude of Washington, D. C., but that farther north greater thickness would probably be needed. It will be noted that the greatest thickness of insulating material was used on top of the hives. This agrees with the practice long employed by many beekeepers and still in use in a great many places. Prior to the experiments of Demuth and Phillips it was not generally known that insulation played such an important part in the winter protection of the colony and the beekeeper attempted only to keep the top of the hive warm. There was a very good reason for this. During the winter the bees give off moisture just as they do at any other season of the year. In breathing they exhale moisture as we do. If you will blow your breath against a cold piece of glass you will notice that the moisture condenses and forms tiny drops of water. In the hive this same process goes on and, as the hive is tightly sealed, there is no way for the moisture to escape. Consequently it condenses and accumulates on the coldest surfaces. If the coldest surface should be the roof we would have actual drops of water falling upon the cluster and causing the bees discomfort, need for additional activity, and perhaps death.

Therefore the old way of packing a hive of bees for winter was to supply an abundance of insulation on top of the hive, and sometimes on the sides and back, and allow the front to remain exposed. By this system the moisture collected on the front or sides of the hive, ran down the walls and evaporated at the entrance. This method, too, gave excellent results for years and it is far preferable to no packing at all. Many beekeepers will be unable to go to the trouble and expense of providing the elaborate packing cases referred to above but it is within the easy possibility of any one to pack a colony at the top. The easiest way to do this is to remove the cover of the hive, lay a piece of burlap directly over the frames and on it place an empty super. The super is then filled with any material which will conserve heat, such as shavings, wheat chaff, cork dust or even leaves. Whatever material is used must be dry. Wet packing material does not insulate because the air spaces in it are more or less filled with water and heat can escape very easily. Such wet material too will freeze in solid mass and become still less useful.

The bees should be packed before the weather becomes too cold in the fall. It has been found that if packing is delayed until cold weather, the bees may be so stimulated by the sudden warmth that they will start rearing brood and thus waste their stores and perish before spring.

The packing should be removed early enough in spring to allow the beekeeper to examine the colonies and to determine their condition. I have already indicated the value of this early spring inspection and it should not be neglected. However, if the weather remains cool the hive should be given some sort of temporary protection.

The value of this spring protection is often overlooked by the beekeeper who has carefully carried his stock over the winter. The colonies are to be protected, especially on those cold raw windy days when the cold "seems to go right through one." It goes "right through" the hives just as it does through your clothing and

the bees need protection at this time more
than at any other because they now have
a host of little ones to keep warm in addition to the wintered-over adults.

I mentioned that there were three factors in successful wintering but thus far
I have talked chiefly of one of them—
shelter. The next important factor is that
of food and on it depends a great deal of
the success one will have in bringing his
bees through a winter. No matter how
perfect the hives or how carefully they
are packed, the bees will stand a good
chance of dying if they do not have the
right kind of food.

It has been the fashion for writers on
apiculture to pass the buck on the subject of winter stores by saying "The bees
should enter the winter with *sufficient*
stores of good quality." I confess that I
wondered for a long time just what they
meant by "sufficient" and finally realized that they did not know themselves—
at least not very definitely.

It is now generally known, however, that the amount of food required by a colony will vary greatly with the colony, with the season, and with the kind of food. It is small wonder that the older writers were not very clear on the subject. Experimentally it has been found that an average colony will consume during the winter months somewhere between fourteen and eighteen pounds of good honey. If fed with cane sugar sirup they will require less than half that amount. Supplied with honey dew they might require more. These figures, too, do not take into consideration the variations in styles of winter packing. A heavily packed hive will house its colony in comfort on much less food than will one that is exposed to the cold.

To accept such experimental figures as fixing a definite maximum consumption in an average winter may lead the beekeeper into trouble, however, because there are sometimes unforeseen conditions that must be met. For instance, last winter I had a colony of Italian bees that was short of stores in the fall. I fed them ten pounds of sugar sirup in addition to what honey they had and felt sure that they would "pull through" all right. Followed a very mild winter with early spring, and brood rearing began at a date unheard of before in our neighborhood. Consequently that colony ate up all of its stores and was on the point of starvation when I discovered its condition.

Therefore I would prefer to have an excess of food in the hive rather than to run the risk of the bees starving late in winter. If honey of good quality is on hand there should be around forty pounds of it. Half that quantity of good sugar sirup would probably be enough.

Sugar sirup is prepared by boiling two parts of sugar to one part of water. To this is added one ounce of tartaric acid to each fifty pounds of sugar to prevent it from crystallizing in the combs. Even

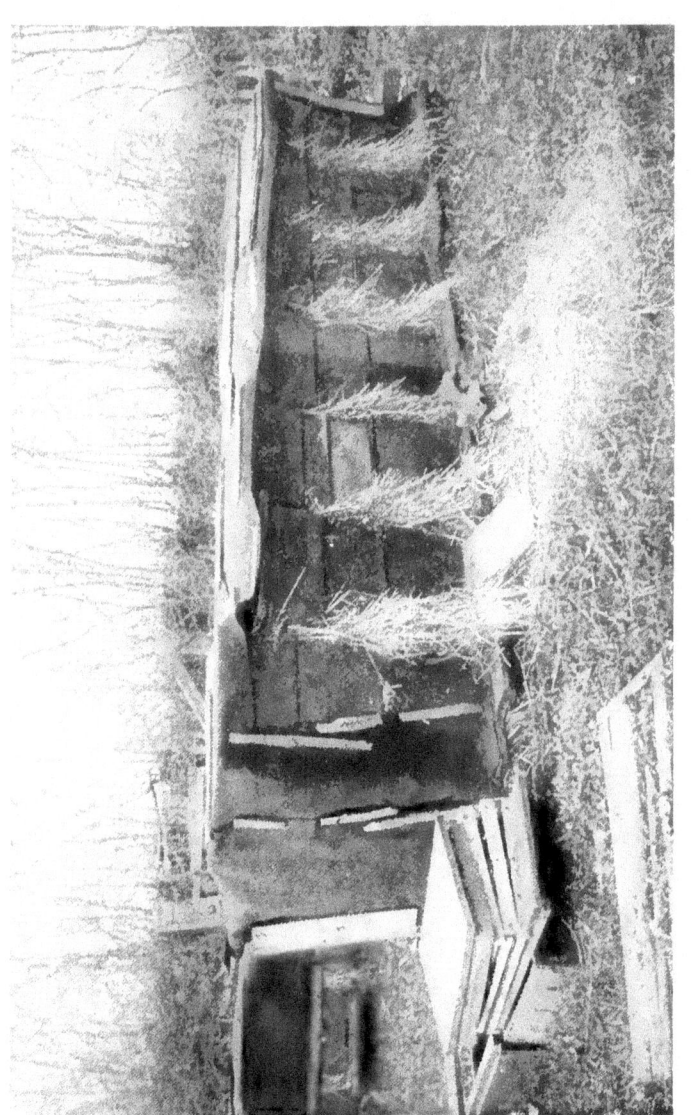

A row of hives packed for winter.

This winter cluster was photographed after they had starved to death—the inevitable result when colonies enter the winter in a weak condition.

though the colony is provided with plenty
of honey it is a good plan to feed them a
few pounds of this sugar sirup just before they are packed. This food is stored
in the space nearest the cluster, it will be
the first used and will assist in keeping
the bees in good condition through the
winter.

Sugar sirup is the best possible winter
food for bees for the reason that it is almost completely consumed within the
body of the bee. There are often weeks
and months at a time when the bees are
unable to fly and as a result they must retain within their intestines the waste matter contained in the food they eat. If
wintered in a warm climate they might
be fed almost anything that was sweet,
but in the North where the winters present a problem, they should be given food
which will be as free from waste as it may
be. Sugar is the ideal food, good white
honey is next in value and dark honey
and honey dew are the poorest possible.

In the report of their experiments with wintering bees, Demuth and Phillips compare a colony with a furnace and the illustration is so good that I will not attempt to improve on their words but will quote it just as it is given. They say: "Let us assume that we have a furnace for heating a building so constructed that ashes may be removed only when the temperature of the outside air is warm. If the house has thin walls and many openings, the furnace can not maintain a high temperature in extreme cold weather, the amount of fuel is increased, the ashes accumulate rapidly and clog the furnace, and in a desperate effort to raise the house temperature we would probably burn out the furnace. On the other hand, if the house is well built and heavily insulated, a low fire will suffice, and as a result there will be a minimum amount of ashes. The better the fuel, the less the amount of ashes in either case.

"It is permissible to compare a colony

of bees as a unit of heat production with
this furnace. If the bees are in a single
walled hive in a cold climate, the colony
must generate a great amount of heat,
must consume much more honey, and
feces will accumulate rapidly. As the
bees are unable to discharge their feces
until the temperature of the outer air is
high enough for flight, the 'furnace' is
clogged. The bees are 'burned out' by
excessive heat production, and, even worse
than in the case of the furnace, the irri-
tation resulting from the presence of
feces causes still more heat production.
On the other hand, if abundantly insulated, the heat generated is conserved, the
consumption of stores and the amount of
feces are reduced, and the bees can readily retain the feces until a flight day, in
any place in which the bees can be kept.
The better the stores the less the amount
of feces in either case. We should not
expect much of a furnace in an open shed,
and we have no more right to expect good

results from a colony wintered in a thin walled hive in a cold climate, or even in a better hive placed in a windy location."

The third factor that I mentioned at the beginning of this chapter was that of numbers—the population of the bee city. A small colony will not have enough bees to form a cold resisting cluster in severe weather. They will bunch up in a compact ball and the number on the inside will not be great enough to generate the needful heat. There will be no surplus of warm bees to crawl out of the cluster and relieve the cold members of the insulating layer. Consequently the entire cluster becomes chilled through and quietly passes on to a bee heaven where there is plenty of white clover and no smoke.

In order that the bees may winter to the best advantage the colony should be strong in numbers at the end of the summer and these bees which are to winter over should be as young as possible. In other words, brood rearing should be encouraged up to the time of frost.

This late brood insures that a majority of the bees for wintering over are young and vigorous, not tired out and travel-worn after a summer's hard work in the fields. Such young bees will not only have a better chance of resisting the cold but they will be far more capable of conducting the work of the hive in a successful manner in the spring. They will have all of that youthful "pep" which goes far toward making a success of any new venture—and each year in the beehive is a new venture for the honey gatherers. All old scores are wiped out, a new generation is started.

Bearing in mind these three factors of shelter, food and numbers, any beekeeper should be able to work out his wintering problems except in those extreme northern locations where cellar wintering must be resorted to. I have not gone into the question of cellar wintering chiefly because it is not widely practised and partly because I never wintered a bee in a cellar in my life. And I dislike to write on any subject on which I lack first-hand information.

The principles of cellar wintering, however, do not differ from those of outside wintering except that more stress must be laid on the factor of food. Bees wintered in a cellar must pass several months without flight. Consequently the food must be of the best quality. The cellar itself furnishes the insulation to the hive and in many ways it provides ideal wintering. It is not a practice for the beginner with bees, however, and in these chapters I have always had in mind the enthusiastic amateur, either actual or prospective, rather than the confirmed victim.

THE END

INDEX

INDEX

www.ingramcontent.com/pod-product-compliance
Lightning Source LLC
Chambersburg PA
CBHW051901170526
45168CB00001B/199